T0181577

Advances in Intelligent Systems and Computing

Volume 555

Series editor

Janusz Kacprzyk, Polish Academy of Sciences, Warsaw, Poland
e-mail: kacprzyk@ibspan.waw.pl

About this Series

The series "Advances in Intelligent Systems and Computing" contains publications on theory, applications, and design methods of Intelligent Systems and Intelligent Computing. Virtually all disciplines such as engineering, natural sciences, computer and information science, ICT, economics, business, e-commerce, environment, healthcare, life science are covered. The list of topics spans all the areas of modern intelligent systems and computing.

The publications within "Advances in Intelligent Systems and Computing" are primarily textbooks and proceedings of important conferences, symposia and congresses. They cover significant recent developments in the field, both of a foundational and applicable character. An important characteristic feature of the series is the short publication time and world-wide distribution. This permits a rapid and broad dissemination of research results.

Advisory Board

Chairman

Nikhil R. Pal, Indian Statistical Institute, Kolkata, India
e-mail: nikhil@isical.ac.in

Members

Rafael Bello Perez, Universidad Central "Marta Abreu" de Las Villas, Santa Clara, Cuba
e-mail: rbellop@uclv.edu.cu

Emilio S. Corchado, University of Salamanca, Salamanca, Spain
e-mail: escorchado@usal.es

Hani Hagras, University of Essex, Colchester, UK
e-mail: hani@essex.ac.uk

László T. Kóczy, Széchenyi István University, Győr, Hungary
e-mail: koczy@sze.hu

Vladik Kreinovich, University of Texas at El Paso, El Paso, USA
e-mail: vladik@utep.edu

Chin-Teng Lin, National Chiao Tung University, Hsinchu, Taiwan
e-mail: ctlin@mail.nctu.edu.tw

Jie Lu, University of Technology, Sydney, Australia
e-mail: Jie.Lu@uts.edu.au

Patricia Melin, Tijuana Institute of Technology, Tijuana, Mexico
e-mail: epmelin@hafsamx.org

Nadia Nedjah, State University of Rio de Janeiro, Rio de Janeiro, Brazil
e-mail: nadia@eng.uerj.br

Ngoc Thanh Nguyen, Wroclaw University of Technology, Wroclaw, Poland
e-mail: Ngoc-Thanh.Nguyen@pwr.edu.pl

Jun Wang, The Chinese University of Hong Kong, Shatin, Hong Kong
e-mail: jwang@mae.cuhk.edu.hk

More information about this series at http://www.springer.com/series/11156

Srikanta Patnaik · Florin Popentiu-Vladicescu
Editors

Recent Developments in Intelligent Computing, Communication and Devices

Proceedings of ICCD 2016

 Springer

Editors
Srikanta Patnaik
Department of Computer Science
  and Engineering, Faculty of Engineering
  and Technology
SOA University
Bhubaneswar, Odisha
India

Florin Popentiu-Vladicescu
Romania Director of UNESCO Chair
University of Oradea
Bucharest, Oradea
Romania

ISSN 2194-5357   ISSN 2194-5365   (electronic)
Advances in Intelligent Systems and Computing
ISBN 978-981-10-3778-8   ISBN 978-981-10-3779-5   (eBook)
DOI 10.1007/978-981-10-3779-5

Library of Congress Control Number: 2017930162

Printed on acid-free paper

This Springer imprint is published by Springer Nature
The registered company is Springer Nature Singapore Pte Ltd.
The registered company address is: 152 Beach Road, #21-01/04 Gateway East, Singapore 189721, Singapore

Preface

Intelligence is the new paradigm for any branch of engineering and technology. *Intelligence* has been defined in many different ways including as one's capacity for logic, understanding, self-awareness, learning, emotional knowledge, planning, creativity and problem solving. Formally speaking, it is described as the ability to perceive information and retain it as knowledge to be applied towards adaptive behaviours within an environment or context.

Artificial intelligence is the study of intelligence in machines, which was coined during the 1950s, and this was a proposition much before to this may be well back to fourth century B.C. Let us take a brief look at the history of intelligence. Aristotle invented syllogistic logic in fourth century B.C., which is the first formal deductive reasoning system. In thirteenth century, in 1206 A.D., Al-Jazari, an Arab inventor, designed what is believed to be the first programmable humanoid robot, a boat carrying four mechanical musicians powered by water flow. In 1456, printing machine using moveable type was invented and Gutenberg Bible was printed. In fifteenth century, clocks were first produced using lathes, which were the first modern measuring machines.

In 1515, clockmakers extended their craft for creating mechanical animals and other novelties. In the early seventeenth century, Descartes proposed that bodies of animals are nothing more than complex machines. Many other seventeenth-century thinkers offered variations and elaborations of Cartesian mechanism. In 1642, Blaise Pascal created the first mechanical digital calculating machine. In 1651, Thomas Hobbes published *"The Leviathan"*, containing a mechanistic and combinatorial theory of thinking. Between 1662 and 1666, Arithmetical machines were devised by Sir Samuel Morland. In 1673, Leibniz improved Pascal's machine to do multiplication and division with a machine called the step reckoner and envisioned a universal calculus of reasoning by which arguments could be decided mechanically. The eighteenth century saw a profusion of von Kempelen's phony mechanical chess player.

In 1801, Joseph-Marie Jacquard invented the Jacquard loom, the first programmable machine, with instructions on punched cards. In 1832, Charles Babbage and Ada Byron designed a programmable mechanical calculating machine, the

Analytical Engine, whose working model was built in 2002. In 1854, George Boole developed a binary algebra representing some "laws of thought". In 1879, modern propositional logic was developed by Gottlob Frege in his work Begriffsschrift and later clarified and expanded by Russell, Tarski, Godel, Church and others.

In the first half of twentieth century, Bertrand Russell and Alfred North Whitehead published Principia Mathematica, which revolutionized formal logic. Russell, Ludwig Wittgenstein and Rudolf Carnap lead philosophy into logical analysis of knowledge. In 1912, Torres Y. Quevedo built his chess machine "Ajedrecista", using electromagnets under the board to play the endgame rook and king against the lone king, the first computer game.

During the second part of twentieth century, the subject was formally taken a shape in the name of traditional artificial intelligence following the principle of physical symbolic system hypothesis to get great success, particularly in knowledge engineering. During the 1980s, Japan proposed the fifth-generation computer system (FGCS), which is knowledge information processing forming the main part of applied artificial intelligence. During the next two decades, key technologies for the FGCS were developed such as VLSI architecture, parallel processing, logic programming, knowledge base system, applied artificial intelligence and pattern processing. The last decade observed the achievements of intelligence in mainstream computer science and at the core of some systems such as communication, devices, embedded systems and natural language processor.

This volume covers some of the recent developments of intelligent sciences in its three tracks, namely intelligent computing, intelligent communication and intelligent devices. Intelligent computing track covers areas such as intelligent and distributed computing, intelligent grid and cloud computing, Internet of Things, soft computing and engineering applications, data mining and knowledge discovery, Semantic and Web Technology, hybrid systems, agent computing, bioinformatics and recommendation systems.

At the same time, intelligent communication covers communication and network technologies, including mobile broadband and all optical networks that are the key to groundbreaking inventions of intelligent communication technologies. This covers communication hardware, software and networked intelligence, mobile technologies, machine-to-machine communication networks, speech and natural language processing, routing techniques and network analytics, wireless ad hoc and sensor networks, communications and information security, signal, image and video processing, network management and traffic engineering.

The intelligent device is any equipment, instrument, or machine that has its own computing capability. As computing technology becomes more advanced and less expensive, it can be built into an increasing number of devices of all kinds. The intelligent device covers areas such as embedded systems, RFID, RF MEMS, VLSI design and electronic devices, analogue and mixed-signal IC design and testing, MEMS and microsystems, solar cells and photonics, nanodevices, single electron and spintronics devices, space electronics and intelligent robotics.

We shall not forget to inform you that the next edition of the conference, i.e. 3rd International Conference on Intelligent Computing, Communication and Devices

(ICCD-2017), is going to be held during June 2017 in China, and we shall be updating you regarding the dates and venue of the conference.

I am sure that the readers shall get immense ideas and knowledge from this volume on recent developments on intelligent computing, communication and devices.

Bhubaneswar, India   Prof. Srikanta Patnaik
Oradea, Romania   Prof. Florin Popentiu-Vladicescu

Acknowledgements

The papers coved in this proceeding are the result of the efforts of the researchers working in this domain. We are thankful to the authors and paper contributors of this volume.

We are thankful to the editor of the Springer Book Series on "**Advances in Intelligent Systems and Computing**" Prof. Janusz Kacprzyk for his support to bring out the second volume of ICCD-2016. It is noteworthy to mention here that this was really a big boost for us to continue this conference series on "International Conference on Intelligent Computing, Communication and Devices", for the second edition.

We shall fail in our duty if we will not mention the role of Mr. Aninda Bose, Senior Editor, Hard Sciences, Springer.

We are thankful to our friend Dr. Florin Popentiu-Vladicescu, Director of UNESCO Chair, University of Oradea, Bucharest, Romania, for his key note address. We are also thankful to the experts and reviewers who have worked for this volume despite of the veil of their anonymity.

About the Book

This book presents high-quality papers presented at 2nd International Conference on Intelligent Computing, Communication and Devices (ICCD 2016) organized by Interscience Institute of Management and Technology (IIMT), Bhubaneswar, Odisha, India, during 13 and 14 August 2016. This book covers all dimensions of intelligent sciences in its three tracks, namely intelligent computing, intelligent communication and intelligent devices. Intelligent computing track covers areas such as intelligent and distributed computing, intelligent grid and cloud computing, Internet of things, soft computing and engineering applications, data mining and knowledge discovery, semantic and Web technology, hybrid systems, agent computing, bioinformatics and recommendation systems.

Intelligent communication covers communication and network technologies, including mobile broadband and all optical networks that are the key to ground-breaking inventions of intelligent communication technologies. This covers communication hardware, software and networked intelligence, mobile technologies, machine-to-machine communication networks, speech and natural language processing, routing techniques and network analytics, wireless ad hoc and sensor networks, communications and information security, signal, image and video processing, network management and traffic engineering.

And finally, the third track intelligent device deals with any equipment, instrument or machine that has its own computing capability. As computing technology becomes more advanced and less expensive, it can be built into an increasing number of devices of all kinds. The intelligent device covers areas such as embedded systems, RFID, RF MEMS, VLSI design and electronic devices, analog and mixed-signal IC design and testing, MEMS and microsystems, solar cells and photonics, nanodevices, single electron and spintronics devices, space electronics and intelligent robotics.

Contents

Research on SaaS-Based Mine Emergency Rescue Preplan and Case Management System 1
Shancheng Tang, Bin Wang, Xinguan Dai and Yunyue Bai

An Investigation of Matching Approaches in Fingerprints Identification 9
Asraful Syifaa' Ahmad, Rohayanti Hassan, Noraini Ibrahim, Mohamad Nazir Ahmad and Rohaizan Ramlan

Figure Plagiarism Detection Using Content-Based Features 17
Taiseer Eisa, Naomie Salim and Salha Alzahrani

An Efficient Content-Based Image Retrieval (CBIR) Using GLCM for Feature Extraction 21
P. Chandana, P. Srinivas Rao, C.H. Satyanarayana, Y. Srinivas and A. Gauthami Latha

Emotion Recognition System Based on Facial Expressions Using SVM 31
Ielaf Osaamah Abdul-Majjed

An AES–CHAOS-Based Hybrid Approach to Encrypt Multiple Images 37
Shelza Suri and Ritu Vijay

Automatic Text Summarization of Video Lectures Using Subtitles 45
Shruti Garg

Classification of EMG Signals Using ANFIS for the Detection of Neuromuscular Disorders 53
Sakuntala Mahapatra, Debasis Mohanta, Prasant Kumar Mohanty and Santanu Kumar Nayak

Evaluating Similarity of Websites Using Genetic Algorithm
for Web Design Reorganisation   61
Jyoti Chaudhary, Arvind K. Sharma and S.C. Jain

Fusion of Misuse Detection with Anomaly Detection
Technique for Novel Hybrid Network Intrusion Detection System   73
Jamal Hussain and Samuel Lalmuanawma

Analysis of Reconfigurable Fabric Architecture
with Cryptographic Application Using Hashing Techniques   89
Manisha Khorgade and Pravin Dakhole

Privacy Preservation of Infrequent Itemsets Mining
Using GA Approach   97
Sunidhi Shrivastava and Punit Kumar Johari

A Quinphone-Based Context-Dependent Acoustic
Modeling for LVCSR   105
Priyanka Sahu and Mohit Dua

Slot-Loaded Microstrip Antenna: A Possible Solution
for Wide Banding and Attaining Low Cross-Polarization   113
Ghosh Abhijyoti, Chakraborty Subhradeep, Ghosh Kumar Sanjay,
Singh L. Lolit Kumar, Chattopadhyay Sudipta and Basu Banani

Fractal PKC-Based Key Management Scheme
for Wireless Sensor Networks   121
Shantala Devi Patil, B.P. Vijayakumar and Kiran Kumari Patil

Histogram-Based Human Segmentation Technique
for Infrared Images   129
Di Wu, Zuofeng Zhou, Hongtao Yang and Jianzhong Cao

Medical Image Segmentation Based on Beta Mixture
Distribution for Effective Identification of Lesions   133
S. Anuradha and C.H. Satyanarayana

Toward Segmentation of Images Based on Non-Normal
Mixture Models Based on Bivariate Skew Distribution   141
Kakollu Vanitha and P. Chandrasekhar Reddy

Development of Video Surveillance System in All-Black
Environment Based on Infrared Laser Light   149
Wen-feng Li, Bo Zhang, M.T.E. Kahn, Meng-yuan Su, Xian-yu Qiu
and Ya-ge Guo

Data Preprocessing Techniques for Research
Performance Analysis   157
Fatin Shahirah Zulkepli, Roliana Ibrahim and Faisal Saeed

Author Index   163

About the Editors

Dr. Srikanta Patnaik is a professor in the Department of Computer Science and Engineering, Faculty of Engineering and Technology, SOA University, Bhubaneswar, India. He has received his Ph.D. (Engineering) on computational intelligence from Jadavpur University, India, in 1999 and supervised 12 Ph.D. theses and more than 30 M.Tech. theses in the area of computational intelligence, soft computing applications and re-engineering. Dr. Patnaik has published around 60 research papers in international journals and conference proceedings. He is author of 2 text books and edited 12 books and few invited chapters in various books, published by leading international publishers such as Springer-Verlag and Kluwer Academic. Dr. Patnaik was the principal investigator of AICTE-sponsored TAPTEC project "Building Cognition for Intelligent Robot" and UGC-sponsored major research project "Machine Learning and Perception using Cognition Methods". He is the editor in chief of International Journal of Information and Communication Technology and International Journal of Computational Vision and Robotics published from Inderscience Publishing House, England, and also series editor of book series on "Modeling and Optimization in Science and Technology" published from Springer, Germany.

Dr. Florin Popentiu-Vladicescu is at present an associated professor of software engineering at UNESCO Department University, City University, London. Dr. Florin Popentiu has been a visiting professor at various universities such as Telecom Paris, ENST, Ecole Nationale Superieure des Mines Paris, ENSMP, Ecole Nationale Superieure de Techniques Avancees, ENSTA, ETH—Zurich, Université Pierre et Marie Curie Paris, UPMC, Delft University of Technology, University of Twente Enschede and Technical University of Denmark Lyngby. Prof. Florin Popentiu-Vladicescu is currently a visiting professor at "ParisTech" which includes the "Grandes Ecoles", The ATHENS Programme, where he teaches courses on software reliability. He also lectures on software reliability at International Master of Science in Computer Systems Engineering, Technical University of Denmark. Prof. Florin Popentiu-Vladicescu has published over 100 papers in international journals and conference proceedings and is author of one book and co-author of

3 books. He has worked for many years on problems associated with software reliability and has been co-director of two NATO research projects involving collaboration with partner institutions throughout Europe. He is on advisory board of several international journals: Reliability: Theory & Applications; Journal of Systemics, Cybernetics and Informatics (JSCI); and Microelectronics Reliability. He is reviewer for ACM Computing Reviews, IJCSIS, and associated editor to IJICT.

Research on SaaS-Based Mine Emergency Rescue Preplan and Case Management System

Shancheng Tang, Bin Wang, Xinguan Dai and Yunyue Bai

Abstract The existing emergency rescue plan and case management systems face the following problems: The relevant systems are maintained by different rescue brigades, but staff in those bridges have limited professional skills to conduct maintenance; the systems are independent to each other, and the data interface is not exactly consistent with the National Safety Supervision Bureau, so that the bureau and provincial rescue centers cannot master the national rescue information timely and accurately. In addition, the rescue brigades cannot share the rescue preplan and cases to reduce the maintenance cost. In this paper, leveraged by the concept of Software as a Service (SaaS), we propose a mine emergency rescue preplan and case management system to realize national-, provincial-, and brigade-level rescue planning and case management, unified interface, entirely shared resources, and reduced maintenance costs. The system test results verify that this proposed platform can easily support all the concurrent national emergency rescue management units/users. The proposed system leveraged by SaaS is 3 times faster than the traditional Java EE-based system: 90% of transactions in the newly proposed system are finished in 650 ms and 50% of those transactions are finished in 32 ms.

Keywords Emergency rescue preplan and case management · Saas · Service-oriented architecture

1 Introduction

"The overall construction plan of the national production safety supervision information platform" contains five business systems: safety production industry regulations, coal mine supervision, comprehensive supervision, the public services, and emergency rescue [1, 2]. Emergency rescue preplan and case management

S. Tang (✉) · B. Wang · X. Dai · Y. Bai
Communication and Information Institute, Xi'an University of Science
and Technology, Xi'an, China
e-mail: tangshancheng@21cn.com

© Springer Nature Singapore Pte Ltd. 2017
S. Patnaik and F. Popentiu-Vladicescu (eds.), *Recent Developments in Intelligent Computing, Communication and Devices*, Advances in Intelligent Systems and Computing 555, DOI 10.1007/978-981-10-3779-5_1

system is an important part of the national production safety supervision information platform, and many associated systems have been proposed recently [3–10]. In addition, effective communication is a critical requirement for the rescuer system, and some solutions are also proposed to increase the communication performance [11–14]. However, the existing emergency rescue preplan and case management systems still face the following problems. First of all, the system belongs to independent rescue brigades, so that the data interface is not exactly consistent with the National Safety Supervision Bureau. Secondly, the bureau cannot communicate effectively with the national rescue center timely and effectively. Thirdly, the rescue brigades cannot share the rescue planning and cases to reduce the cost. Last but not least, the rescue brigades who have no professional skill to maintain the systems need dedicated servers and personnel to maintain the existing system, resulting in a waste of resources. To solve the above problems, we put forward the mine emergency rescue preplan and case management system based on Software as a Service (SaaS). The system test results verify that this proposed platform can easily support all the concurrent national emergency rescue management units/users. Additionally, the communication performance has been greatly enhanced. Specifically, the proposed system is 3 times faster than the traditional Java EE-based system: 90% of transactions in the newly proposed system are finished in 650 ms and 50% of those transactions are finished in 32 ms.

2 SaaS and Service-Oriented Architecture

There are three cloud computing service models: IaaS, PaaS, and SaaS. Software as a Service (SaaS) is a new software infrastructure construction method and system service software delivery model in which software is centralized, managed, and maintained. Users do not need to manage software, but they can use software in real time. SaaS is also called "on-demand software." In the civil and military information system construction, SaaS gets more and more attention, including management system, automation office software, communication software, and database management software. SaaS covers almost all areas [15, 16].

Service-oriented architecture (SOA) is a service architecture which is coupled loosely and has coarse grain size services. In SOA, services interact with each other through simple interface which has nothing to do with programming languages and communication protocols. SOA has the following important distinguishing features: services, interoperability, and loose coupling. SOA will be able to help software engineers to understand, develop, and deploy systems with various enterprise components [17–20].

3   The Architecture of System

The architecture of system is shown in Fig. 1.

IaaS: Cloud infrastructure services are self-service models for accessing, monitoring, and controlling the infrastructure of remote data centers such as computing (virtualized or bare metal), memory, storage, and network services. All resources are virtualized into services such as computing services, storage services, load management services, and backup services. PaaS: Cloud platform services are provided for applications to serve other cloud components. Developers can take advantage of PaaS as a framework to build customized applications. For PaaS services have a whole set of service logic, it makes the development, testing, and deployment for applications more simple, quick, and highly active. SaaS: Cloud application services are the largest cloud market and are growing rapidly. SaaS utilizes the HTTP to transfer applications that are supervised by third-party vendors, and the client accesses the interface of the applications [15, 16]. Access layer: This

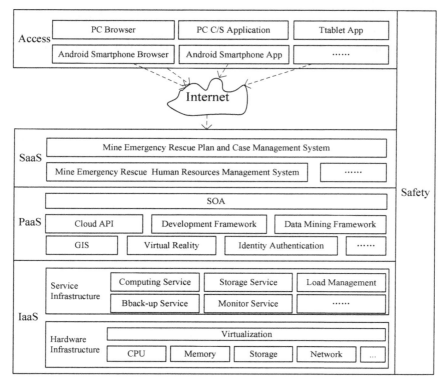

Fig. 1 Architecture of mine emergency rescue preplan and case management system based on SaaS

layer access application provided by SaaS includes variety of terminals Web browser, desktop applications, and smartphone applications. The system can support variety of terminal access with one platform by Internet.

4 The Components of System

The components of system are shown in Fig. 2.

Regional mine subsystem includes the following modules: regional mine information management, ventilation network graph management, underground tunnel map management, and traffic map management. This subsystem is the basis for other subsystems, which is based on GIS, 3D virtual reality services to show regional mines, cases, preplan, and rescue plan.

Preplan expert subsystem includes knowledge base management, inference engine management, interpreter, integrated database, knowledge acquisition, and prescue plan. This subsystem support users to create preplan automatically with expert knowledge.

Preplan management subsystem includes the following modules: coal and gas outburst accident emergency rescue preplan management, mine fire accident rescue preplan management, mine flooding accident rescue preplan management, roof

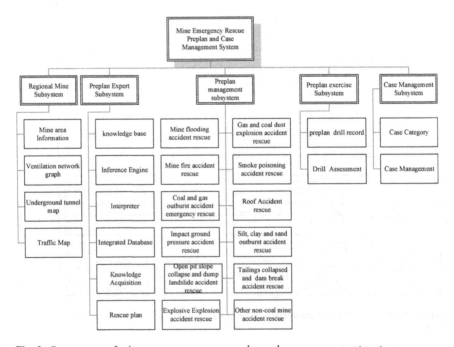

Fig. 2 Components of mine emergency rescue preplan and case management system

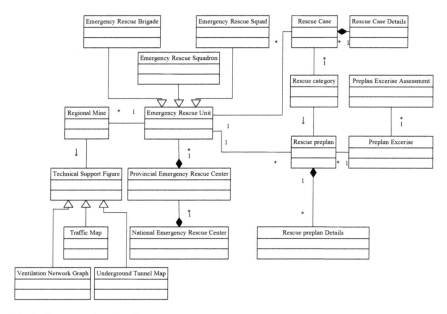

Fig. 3 System entity class diagram

accident rescue preplan management, gas and coal dust explosion accident rescue preplan management, mine flooding accident rescue preplan management, smoke poisoning accident rescue preplan management, silt clay and sand outburst accident rescue preplan management, tailings collapsed and dam break accident rescue preplan management, explosive explosion accident rescue preplan management, and other non-coal mine accident rescue preplan management. This subsystem is the central section. All of the preplans are showed by GIS, 3D virtual reality technology.

Preplan exercise and case management subsystem includes preplan exercise management, preplan exercise assessment management, case category management, and case management.

Through the analysis of the system, we got the system entity class diagram (Fig. 3), and the system entity class diagram is the core concept of the system modeling.

5 System Implementation and Testing

In order to compare the difference between the system based on SaaS and the system based on traditional technology such as Java EE, we implemented the system and deployed on two platforms (one is OpenStack, and the other is Java

Table 1 Test aggregation results (A)

Label	Samples	Average (ms)	Median (ms)	90% line (ms)	95% line (ms)	99% line (ms)
SaaS	3000	213	32	650	1165	2317
Java EE	3000	823	710	1653	1872	3247

Table 2 Test aggregation results (B)

Label	Min (ms)	Max (ms)	Error %	Throughput (transactions/s)	KB/s
SaaS	5	6410	0.00	159.9	12331
Java EE	23	9741	0.00	95.9	609.7

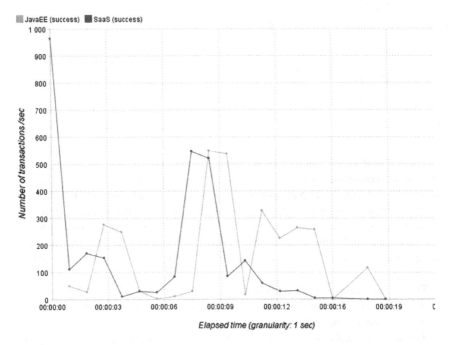

Fig. 4 Transactions per second

EE), and the two platforms have same hardware configurations (6 servers that have same configurations, CPU: Intel Xeon E5645, 6 core 2.4 GHz; 16 G memory).

The system provides about 50 Web service interfaces, based on the JSON data encapsulation. We tested two platforms based on JMeter, as shown in Tables 1 and 2 and Figs. 4 and 5. The system based on SaaS is 3 times (average response time) faster than the system based on Java EE. The throughput (transactions/s) supported by the system based on SaaS is nearly 2 times that of the system based on Java EE. In the system based on SaaS, 90% of transactions are finished in 650 ms and 50% of transactions are finished in 32 ms.

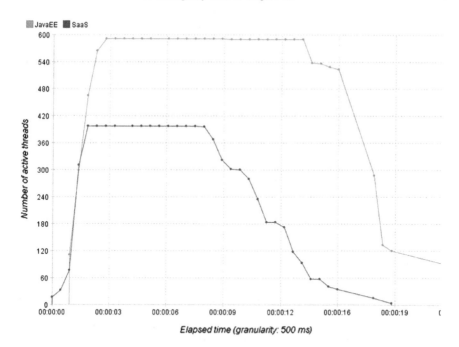

Fig. 5 Active threads over time

6   Conclusion

In order to solve the problems of current mine emergency rescue preplan and case management systems, we propose a new system based on SaaS to realize national-, provincial-, and brigade-level rescue planning, case management, unified interface, entirely shared resources, and reduced maintenance costs. After designing and implementing the system, extensive tests have been conducted to evaluate the advantages of the system. The system test results verify that this proposed platform can easily support all the concurrent national emergency rescue management units/users. Additionally, the communication performance has been greatly enhanced. Specifically, the proposed system is 3 times faster than the traditional Java EE-based system: 90% of transactions in the newly proposed system are finished in 650 ms and 50% of those transactions are finished in 32 ms.

Acknowledgements   This article is sponsored by "Scientific Research Program Funded by Shaanxi Provincial Education Commission (Program NO. 2013JK1079)," "Scientific Research Program Funded by Shaanxi Provincial Education Commission (Program NO. 16JK1501)," "Shaanxi Province Science and Technology Innovation Project (2015KTCQ03-10)," "Xi'an City Research Collaborative Innovation Program (CXY1519 (5))," and "Xi'an Beilin District Science and Technology Project (Gx1601)."

References

1. National security supervision bureau. About print and distribute "the general construction plan of the national production safety supervision information platform" notification. State administration of production safety supervision and administration of the state administration of coal mine safety announcement. 2015.1.20.
2. National security supervision bureau. To promote the preparation of contingency plans from the "have" to "excellent". China Safety Production Report. 2013, Sixth Edition: 1–2.
3. BU Chuangli, PAN Lihu, ZHI Yu. Design of emergency plan management system of coal mine based on case-based seasoning. Industry and Mine Automation. 2015, 41(7): 34–38.
4. GONG Si-han. Design and Implementation of Emergency Plan Management System Based on Workflow. COMPUTER ENGINEERING & SOFTWARE. 2015, 36(11): 89–91.
5. HAO Chuan-bo, LIU Zhen-wen. On Evaluation of Coal Mine Accident Emergency Rescue Plan. Value Engineering. 2014, 3(11): 13–14.
6. Yuan Jian. Establishment and management of Dafosi mine emergency rescue plan. Master's degree thesis of Xi'an University of Science And Technology. 2013.
7. Guo Wen. The Research of Coal Mine Production Safety Emergency Rescue Preplan. Master's degree thesis of Lanzhou University. 2015.
8. Li Zhiliang. A Preliminary Study on Water Disaster Emergency Rescue Counterplan in Coal Mine. China High-Tech Enterprises. 2014, 102–103.
9. Wang Shuai. An Emergency Rescue Plan of Coal Mine Major Accident. Inner Mongolia Coal Economy. 2016, 83–84.
10. Zhang Jun-bo, Guo De-yong, Wang Li-bing. The research of coal mine emergency rescue organization structure model. Journal of China Coal Society. 2012, 37(4): 664–668.
11. S. Wen, W. Fei, and D. Jianbo, An Emergency Communication System based on WMN in underground mine, Computer Application and System Modeling (ICCASM), 2010 International Conference on, vol. 4, no., pp. V4-624–V4-627, 22–24 Oct. 2010.
12. G. Wang, Y. Wu, K. Dou, Y. Ren, and J. Li, AppTCP: The design and evaluation of application-based TCP for e-VLBI in fast long distance networks, Future Generation Computer Systems, vol. 39, pp. 67–74, 2014.
13. G. Wang, Y. Ren, K. Dou, and J. Li, IDTCP: An effective approach to mitigating the TCP Incast problem in data center networks, Information Systems Frontiers, vol. 16, pp. 35–44, 2014.
14. G. Wang, Y. Ren, and J. Li, An effective approach to alleviating the challenges of transmission control protocol, IET Communications, vol. 8, no. 6, pp. 860–869, 2014.
15. Paul Gil. What Is 'SaaS' (Software as a Service). http://netforbeginners.about.com/od/s/f/what_is_SaaS_software_as_a_service.htm.
16. Ziff Davis. Definition of: SaaS. PC Magazine Encyclopedia. Retrieved 14 May 2014.
17. Zhao Hui-qun, Sun Jing. A Methodological Study of Evaluating the Dependability of SOA Software System. Journal of computer. 2010, 33(11): 2202–2210.
18. Deng Fan, Chen Ping, Zhang Li-yong, Li Sun-de. Study on distributed policy evaluation engine in SOA environment. Journal of huazhong university of science and technology (natural science edition). 2014, 42(12): 106–110.
19. Tan Wei, Dong Shou-bin, Liu Xuan. Research on self-adaptive software for variable business process based on semantic SOA. Journal of guangxi university (natural science edition), 2014, 39 (5) : 1123–1130.
20. Zhang Chunxia, Li Xudong, Xu Tao. Discussion about the Core Ideas of Service-Oriented Architecture. Computer Systems & Applications, 2010, 19(6): 251–256.

An Investigation of Matching Approaches in Fingerprints Identification

**Asraful Syifaa' Ahmad, Rohayanti Hassan, Noraini Ibrahim,
Mohamad Nazir Ahmad and Rohaizan Ramlan**

Abstract Fingerprints identification is one of the most widely used biometric technologies that can enhance the security for an access to a system. It is known as the most reliable application compared to others. In the framework of fingerprints identification, the most crucial step is the matching phase. Thus, this paper is devoted to identify and review the existing matching approaches in the specialized literature. The literatures that related to the fingerprints matching were searched using all the relevant keywords. Thirty-five studies were selected as primary sources which comprised of 34 journal articles and a book. The overview of the generic processes was provided for each fingerprints matching. Besides, current works for each of the approaches were addressed according to the issues being handled.

Keywords Biometrics · Fingerprints identification · Correlation based · Minutiae based · Ridge feature based

A.S. Ahmad (✉) · R. Hassan · N. Ibrahim · M.N. Ahmad
Faculty of Computing, Universiti Teknologi Malaysia,
81310 Johor Bharu, Malaysia
e-mail: asrafulsyifaa.ahmad@gmail.com

R. Hassan
e-mail: rohayanti@utm.my

N. Ibrahim
e-mail: noraini_ib@utm.my

M.N. Ahmad
e-mail: mnazir@utm.my

R. Ramlan
Faculty of Technology and Business Management, Universiti Tun Hussein Onn,
86400 Parit Raja, Batu Pahat, Johor, Malaysia
e-mail: rohaizan@uthm.edu.my

© Springer Nature Singapore Pte Ltd. 2017
S. Patnaik and F. Popentiu-Vladicescu (eds.), *Recent Developments in Intelligent
Computing, Communication and Devices*, Advances in Intelligent Systems
and Computing 555, DOI 10.1007/978-981-10-3779-5_2

1   Introduction

Present-day, old security methods that used access card and password are not
excellent enough to protect individuals' belongings. Therefore, biometric authentication systems had been introduced to overcome the limitation of the existing
methods, as they are unique and cannot be stolen. For example, among the traits
from human body part that can be used for recognition process are DNA, iris,
retina, face, voice, signature, palm print, hand geometry, and hand vein [1].
Primarily, biometric authentication process collects all the biometrics information
and keeps them in a database for verification. However, the procedure was claimed
unsecured. Thus, the current methods enroll the biometric template with the newly
produced template by using algorithm or mathematical calculations. The aim was to
find the similarities between both the templates, and the access will be granted if the
pairs match [2].

The key factors of fingerprints usage are the uniqueness of its pattern [3] and the
acceptability to people as a comparison trait compared to other biometric modals.
Fingerprints are one of the most widely used metrics for identification among all
biometrics [4, 5]. The advanced technology of this identification technique had been
deeply explored by the law enforcement agencies and positively used for security
purposes that include getting access to a control system, passing the country border,
and in criminal investigation. Fingerprints identification or dactyloscopy compares
two instances of minutiae from human fingers to determine whether both are from
the same individual. The uniqueness of fingerprints are the distinct pattern of a
series of delta, ridge ending, furrows, and also the characteristic of local ridges.

This paper is organized as follows: Sect. 2 explains the research method used,
Sect. 3 describes the fingerprint matching approaches in detail, and lastly, Sect. 4
reviewed the current issues related to approaches.

2   Research Method

Following the recommendation by Achimugu et al. [6], the review protocols consist
of four main phases as follows: map out the research questions, designing the search
strategy, study the evaluation results, and lastly interpretation and synthesis of data.
The subsequent strategy search designed phases are search terms and resources and
search process, while the study evaluation of subsequent phase included scrutiny
and assessment of quality criteria.

2.1 Research Questions

The aim of this review was to understand and summarize the properties of matching approaches used in fingerprints identification and to identify the possible area for further research in order to complement the performance of existing technique. The following three research questions were simultaneously explored and interlaced to achieve this aim:

- What are the existing techniques in matching approaches used for fingerprints identification?
- What are the generic process and limitations for matching approaches?
- What are the current issues in fingerprint matching?

2.2 Research Strategies

The search strategies used in this research that included the searching by terms and resources were explained in this section. Specifically, the search terms focused on the selected domain of the research and were created by using the following steps [6]:

(1) Derive the major terms from the research questions and identify the synonyms.
(2) Integrate the alternative spellings and synonyms, using the Boolean.
(3) Link the major term using the Boolean AND.

Four mega electronic resources were used as main references including Springer, IEEE Explorer, Science Direct, and Google Scholar. For published journal, papers, title, abstract, and conclusion were used as the parameters. Thereafter, 175 potential studies were realized. By using study selection and scrutiny, only 24 papers were able to provide the answers to the research questions formulated.

3 Fingerprints Matching Approaches

3.1 Taxonomy of the Study

The large numbers of fingerprints matching approaches have been mainly classified into three types which are minutiae based, ridge feature based, and correlation based. Three main taxonomies of fingerprints matching have been shown in Fig. 1.

The correlation-based matching calculates different alignments using the correlation between the corresponding pixels of two superimposed fingerprints images and need to be applied to all possible alignment as the rotation and also displacement are unknown. The minutiae-based matching finds the minutiae alignment

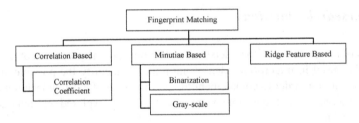

Fig. 1 Taxonomy of fingerprint matching study

between the two fingerprints to find the maximum number of similarities. Another type of matching is ridge feature based that is also known as non-minutiae feature based. The difference of this approach is the usage of feature extracted from the ridge pattern.

3.2 Generic Process of Three Main Types of Fingerprints Matching

Fingerprints matching is a crucial step in both identification and recognition [7]. Basically, fingerprints matching techniques compare and calculate the similarity score between both the input and the template image. Both fingerprints images are called genuine if they obtain a high similarity score and called as impostor if the images were different. The most apparent differences between these three approaches are the input parameter and also the algorithm.

Correlation-based matching used the entire fingerprints structure as its input and all the possible alignments are need to be compared in order to obtain a high matching score [8]. Then, both images are rotated using estimated rotation [9], followed by applying image transform technique to correlate both the template and the input fingerprints images at different rotational and translational alignments before calculating the matching score.

Minutiae-based matching is the most popular approach among others as it includes finding the best alignment in between the image template and the input. In this approach, brute-force algorithm is used in order to find all the possible similar minutiae [10]. After the minutiae extraction phase, the coordinate type of minutiae (ridge ends or bifurcation) is determined. A track of 10 pixels wide is created [11], while any minutiae located in that track are recorded to calculate the matching score.

Ridge feature-based matching utilized the characteristic of fingerprints which are ridges. Local ridge frequency and also local ridge orientation are used to create a set of features that symbolize the individuals' fingerprints [8].

4 Current Issues of the Matching Approaches

This section presents and compares the related work of existing fingerprints matching approaches according to their issue being handled (Table 1). Fingerprints computing can be used to solve noisy and low-quality image, while robustness issues were comprised of the accuracy and precision of the techniques.

Table 1 Comparison of fingerprint matching approaches

Issues	Correlation based	Minutiae based	Ridge feature based
Low-quality image	• Apply correlation coefficient on low-quality input image yet achieve higher matching result [12] • Manipulate the phase-only correlation function that comprises of both global and local search. Each search has its own aim to ensure the matching success [13]	• Implement global distortion correction as preprocessing step that considering texture characteristic of fingerprints image [14] • Apply NCC method for low-quality images and implement normalized cross-correlation method in matching phase [15] • Put different weights on reliable and unreliable minutiae points to calculate similarity score [16]	• Apply this method to find the initial minutiae pair. First implement the ridge matching process. Finally, calculate the matching score [17] • Manipulate hierarchical matching method that employs Level 3 features. The methods performed well and being test on both the high-quality and low-quality images [18]
Robustness	• Apply correlation mapping and discriminative–generative classifier scheme to decide either the input fingerprint is from live person or not. The scheme is based on GMM, Gaussian copula, SVM, and QDA [19] • Design three-step matching phase using the correlation coefficient. First is alignment, next is extraction of common region, and last is calculation of similarity degree [20]	• Included nearest-neighbor matching strategy using different condition of fingerprint image as follows; unaligned, aligned, partial distortion, and occlusion fingerprints [21] • The confidence level of minutiae pair based on the consistency of other pair is considered [22]	• Apply additional dynamic anisotropic pore model (DAPM) pore extraction method to increase the confidence of the matching [23] • Incorporate both minutiae information and ridge features to obtain the similarity score. It also defeats distortion problems in referenced method [24]

Acknowledgements This research is funded by GUP Grant and Universiti Teknologi Malaysia under Vote No: 11H84.

References

1. Jain, Anil K., Arun Ross, and Salil Prabhakar. "An introduction to biometric recognition." IEEE Transactions on circuits and systems for video technology 14, no. 1 (2004): 4–20.
2. Sim, Hiew Moi, Hishammuddin Asmuni, Rohayanti Hassan, and Razib M. Othman. "Multimodal biometrics: Weighted score level fusion based on non-ideal iris and face images." Expert Systems with Applications 41, no. 11 (2014): 5390–5404.
3. Tiwari, Kamlesh, Vandana Dixit Kaushik, and Phalguni Gupta. "An Adaptive Multi-algorithm Ensemble for Fingerprint Matching." In International Conference on Intelligent Computing, pp. 49–60. Springer International Publishing, 2016.
4. Yao, Zhigang, Jean-Marie Le Bars, Christophe Charrier, and Christophe Rosenberger. "A Literature Review of Fingerprint Quality Assessment and Its Evaluation." IET journal on Biometrics (2016).
5. Maltoni, Davide, Dario Maio, Anil Jain, and Salil Prabhakar. Handbook of fingerprint recognition. Springer Science & Business Media, 2009.
6. Achimugu, Philip, Ali Selamat, Roliana Ibrahim, and Mohd Naz'ri Mahrin. "A systematic literature review of software requirements prioritization research." Information and Software Technology 56, no. 6 (2014): 568–585.
7. Peralta, Daniel, Mikel Galar, Isaac Triguero, Daniel Paternain, Salvador García, Edurne Barrenechea, José M. Benítez, Humberto Bustince, and Francisco Herrera. "A survey on fingerprint minutiae-based local matching for verification and identification: Taxonomy and experimental evaluation." Information Sciences 315 (2015): 67–87.
8. Hämmerle-Uhl, Jutta, Michael Pober, and Andreas Uhl. "Towards a Standardised Testsuite to Assess Fingerprint Matching Robustness: The StirMark Toolkit–Cross-Feature Type Comparisons." In IFIP International Conference on Communications and Multimedia Security, pp. 3–17. Springer Berlin Heidelberg, 2013.
9. Nandakumar, Karthik, and Anil K. Jain. "Local Correlation-based Fingerprint Matching." In ICVGIP, pp. 503–508. 2004.
10. Więcław, Łukasz. "A minutiae-based matching algorithms in fingerprint recognition systems." Journal of Medical Informatics & Technologies 13 (2009).
11. Saleh, Amira, A. Wahdan, and Ayman Bahaa. Fingerprint recognition. INTECH Open Access Publisher, 2011.
12. Kumar, S. Sankar, and S. Vasuki. "Performance of Correlation based Fingerprint verification in Real Time." (2016).
13. Shabrina, Nabilah, Tsuyoshi Isshiki, and Hiroaki Kunieda. "Fingerprint authentication on touch sensor using Phase-Only Correlation method." In 2016 7th International Conference of Information and Communication Technology for Embedded Systems (IC-ICTES), pp. 85–89. IEEE, 2016.
14. Moolla, Yaseen, Ann Singh, Ebrahim Saith, and Sharat Akhoury. "Fingerprint Matching with Optical Coherence Tomography." In International Symposium on Visual Computing, pp. 237–247. Springer International Publishing, 2015.
15. Singh, Vedpal, and Irraivan Elamvazuthi. "Fingerprint matching algorithm for poor quality images." The Journal of Engineering 1, no. 1 (2015).
16. Chen, Jiansheng, Fai Chan, and Yiu-Sang Moon. "Fingerprint matching with minutiae quality score." In International Conference on Biometrics, pp. 663–672. Springer Berlin Heidelberg, 2007.
17. Feng, Jianjiang, Zhengyu Ouyang, and Anni Cai. "Fingerprint matching using ridges." Pattern Recognition 39, no. 11 (2006): 2131–2140.

18. Jain, Anil K., Yi Chen, and Meltem Demirkus. "Pores and ridges: High-resolution fingerprint matching using level 3 features." IEEE Transactions on Pattern Analysis and Machine Intelligence 29, no. 1 (2007): 15–27.
19. Akhtar, Zahid, Christian Micheloni, and Gian Luca Foresti. "Correlation based fingerprint liveness detection." In 2015 International Conference on Biometrics (ICB), pp. 305–310. IEEE, 2015.
20. Zanganeh, Omid, Bala Srinivasan, and Nandita Bhattacharjee. "Partial fingerprint matching through region-based similarity." In Digital Image Computing: Techniques and Applications (DICTA), 2014 International Conference on, pp. 1–8. IEEE, 2014.
21. Hany, Umma, and Lutfa Akter. "Speeded-Up Robust Feature extraction and matching for fingerprint recognition." In Electrical Engineering and Information Communication Technology (ICEEICT), 2015 International Conference on, pp. 1–7. IEEE, 2015.
22. Cappelli, Raffaele, Matteo Ferrara, and Davide Maltoni. "Minutiae-based fingerprint matching." In Cross Disciplinary Biometric Systems, pp. 117–150. Springer Berlin Heidelberg, 2012.
23. de Assis Angeloni, Marcus, and Aparecido Nilceu Marana. "Improving the Ridge Based Fingerprint Recognition Method Using Sweat Pores." In Proceedings of the Seventh International Conference on Digital Society. 2013.
24. Liao, Chu-Chiao, and Ching-Te Chiu. "Fingerprint recognition with ridge features and minutiae on distortion." In 2016 IEEE International Conference on Acoustics, Speech and Signal Processing (ICASSP), pp. 2109–2113. IEEE, 2016.

Figure Plagiarism Detection Using Content-Based Features

Taiseer Eisa, Naomie Salim and Salha Alzahrani

Abstract Plagiarism is the process of copying someone else's text or figure verbatim or without due recognition of the source. A lot of techniques have been proposed for detecting plagiarism in texts, but a few techniques exist for detecting figure plagiarism. This paper focuses on detecting plagiarism in scientific figures. Existing techniques are not applicable to figures. Detecting plagiarism in figures requires extraction of information from its components to enable comparison between figures. Consequently, content-based figure plagiarism detection technique is proposed and evaluated based on the existing limitations. The proposed technique was based on the feature extraction and similarity computation methods. Feature extraction method is capable of extracting contextual features of figures in aid of understanding the components contained in figures, while similarity detection method is capable of categorizing a figure either as plagiarized or as non-plagiarized depending on the threshold value. Empirical results showed that the proposed technique was accurate and scalable.

Keywords Figure plagiarism detection · Content feature · Similarity detection

1 Introduction

Roig [9] defines plagiarism as the appropriation of an idea (an explanation, a theory, a result, a conclusion, a hypothesis, or a metaphor) in part or in whole, or with superficial modifications or without giving credit to its originator. There are two types of plagiarism in scientific writing which are plagiarism of text and plagiarism of data. Plagiarism of text takes place when an individual copies a small or large portion of

T. Eisa (✉) · N. Salim
Faculty of Computing, Universiti Teknologi Malaysia, Skudai, Johor, Malaysia
e-mail: taiseralfadil@hotmail.com

S. Alzahrani
Department of Computer Science, Taif University, Taif, Saudi Arabia
e-mail: s.zahrani@tu.edu.sa

© Springer Nature Singapore Pte Ltd. 2017
S. Patnaik and F. Popentiu-Vladicescu (eds.), *Recent Developments in Intelligent Computing, Communication and Devices*, Advances in Intelligent Systems and Computing 555, DOI 10.1007/978-981-10-3779-5_3

text, inserts it into his/her work with or without modifications, and credits citation, while plagiarism of data takes place when an individual copies data as contents of figures or results, inserts it into his/her work with or without modifications, and credits citation. Figures in scientific articles play an important role in disseminating important ideas and describing process and findings which help the readers to understand the concept of existing or published work [2, 4, 5]. Figures and their related textual descriptions account for as much as 50% of a whole paper [3]. Existing techniques are unqualified for detecting plagiarized figures [4, 10, 11]. Plagiarism detection tools discard the figures checking for plagiarism, resulted people can plagiarize diagrams and figures easily without the current plagiarism detection tools detecting it [4, 6].

Scientific figure can be stolen as a whole or as a part, with or without modifications. Modifications in figure can be for the figure's text or for both text and structure. Few methods are proposed to detect figure plagiarism. Arrish et al. [1] proposed a system that relies on changes in the figure shapes and ignoring other features. Analysis method proposed by Rabiu and Salim [8] using structure analysis considers the text and the structure. Plagiarism detection in scientific figure is different from that of normal text and thus requires a different representations and features to capture the plagiarism if occurs. This paper investigates the use of contextual features to detect the plagiarized figures. In this technique, all components in figure are represented using its graphical information. This information is organized in a structural manner with the aim of detecting plagiarism by mapping these features between figures.

2 Content-Based Feature Extraction

Content-based feature extraction is proposed to extract the graphical information in the figures such as textual and structural features. Textual features describe text within the figure. Structural features deals with shapes and flow of components. Image processing and computer vision techniques were used to extract the structural features, while optical character recognition (OCR) were used to extract textual features. The extracted features are saved for further processing in the next steps. Figure 1 shows an example of the content-based feature extraction.

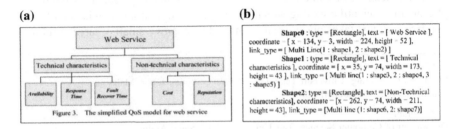

Fig. 1 Results from the content-based extractor of shapes and flow: **a** image of the figure and **b** extracted features

3 Similarity Detection

Content-based figure plagiarism detection method was conducted to measure how much the suspicious figure was similar to that of the source one. Based on the detailed comparisons between the contents of the figures, components in the suspected figures are compared with each component in the source figure according to the attributes of the component. Component-based similarity analysis gives detailed investigation and deep analysis between the suspicious and source figures. Three features were selected to identify the components (shape, text, and flow) due to the significance of those features in a figure. The geometric type of the shape helps to discover plagiarism of change or converted shapes. Similar text gives indicator to similar figure, and the flow between components describes the relationship among components and helps in understanding the sequences of the processes inside the figure. The proposed technique focused on the comparison between the arguments of the suspected figure with similar arguments in the source figure. The overlapping distance between two components was suggested using Jaccard similarity measure. An improvement over existing works which only compare the numbers and types of shapes inside figures such as work of [1]. Similarity between two components is calculated as summation of the arguments' similarities divided by the total number of arguments, while the overall similarities between two figures are calculated as the summation of the components' similarities normalized by the total number of suspicious figure components. Based on the value of overall similarity score of suspicious figure, the figures are classified into plagiarized figure and non-plagiarized, if the value of overall similarity passed the threshold value.

4 Result and Evaluation

Our experiment considered the number of detected plagiarized figures from the source figures. The experiment was performed on 1147 figures (745 source figures, 370 plagiarized figures, and 32 plagiarism-free). The plagiarized figures have different types of modification with different degrees, and some suspicious figures are exact copy, text modification, or text plus structure modifications. Recall, precision and F-measure testing parameters that are commonly used in the information retrieval field are adopted [7]. The results are promising (recall = 0.8 and F-measure 0.9).

5 Conclusions

This paper discussed a new representation method for figure plagiarism detection called content-based representation which represents each component of a figure and provides information about the type of shape, the text inside it, and the relationships with other component(s) to capture the meaning of the figure. The proposed method detects plagiarism based on the figure contents using component-based comparison. This way of comparisons gives more details and can help to discover in detail the difference between the submitted figure (suspicious) and the source figure. The proposed method focused on solving the copy and pasted figure plagiarism, changing text or structure, and deleting or adding words with or without structure modifications.

Acknowledgements This work is supported by the Malaysian Ministry of Higher Education and the Research Management Centre at the Universiti Teknologi Malaysia under Research University Grant Category Vot: Q.J130000.2528.13H46.

References

1. Arrish, S., Afif, F. N., Maidorawa, A., & Salim, N. (2014). Shape-Based Plagiarism Detection for Flowchart Figures in Texts. *International Journal of Computer Science & Information Technology (IJCSIT)* 6(1), 113–124, doi:10.5121/ijcsit.2014.6108.
2. Bhatia, S., & Mitra, P. (2012). Summarizing figures, tables, and algorithms in scientific publications to augment search results. *ACM Trans. Inf. Syst., 30*(1), 1–24, doi:10.1145/2094072.2094075.
3. Futrelle, R. P. Handling figures in document summarization. In *Proc. of the ACL-04 Workshop: Text Summarization Branches Out, Barcelona, Spain, 25–26 July 2004* (pp. 61–65).
4. Hiremath, S., & Otari, M. (2014). Plagiarism Detection-Different Methods and Their Analysis: Review. *International Journal of Innovative Research in Advanced Engineering (IJIRAE), 1*(7), 41–47.
5. Lee, P.-s., West, J. D., & Howe, B. (2016). Viziometrics: Analyzing visual information in the scientific literature. *arXiv preprint* arXiv:1605.04951.
6. Ovhal, P. M., & Phulpagar, B. D. (2015). Plagiarized Image Detection System based on CBIR. *International Journal of Emerging Trends & Technology in Computer Science, 4*(3).
7. Potthast, M., Stein, B., Barr, A., #243, n-Cede, #241, et al. (2010). *An evaluation framework for plagiarism detection.* Paper presented at the Proceedings of the 23rd International Conference on Computational Linguistics: Posters, Beijing, China.
8. Rabiu, I., & Salim, N. (2014). Textual and Structural Approaches to Detecting Figure Plagiarism in Scientific Publications. *Journal of Theoretical and Applied Information Technology, 70*(2), 356–371.
9. Roig, M. (2006). Avoiding plagiarism, self-plagiarism, and other questionable writing practices: a guide to ethical writing.
10. Zhang, X.-x., Huo, Z.-l., & Zhang, Y.-h. (2014). Detecting and (Not) Dealing with Plagiarism in an Engineering Paper: Beyond CrossCheck—A Case Study. [journal article]. *Science and Engineering Ethics, 20*(2), 433–443, doi:10.1007/s11948-013-9460-5.
11. Zhang, Y.-h. H., Jia, X.-y., Lin, H.-f., & Tan, X.-f. (2013). Be careful! Avoiding duplication: a case study. *Journal of Zhejiang University. Science. B, 14*(4), 355.

An Efficient Content-Based Image Retrieval (CBIR) Using GLCM for Feature Extraction

P. Chandana, P. Srinivas Rao, C.H. Satyanarayana, Y. Srinivas and A. Gauthami Latha

Abstract Today, modern technology led to a faster growth of digital media collection, and it contains both still images and videos. Storage devices contain large amount of digital images, increasing the response time of a system to retrieve images required from such collections, which degrades the performance. Various search skills are needed to find what we are searching for in such large collections. The annotations are given manually for images by describing with the set of keywords. By doing so, the contents of an image retrieve images of interest, but it is time-consuming. Also, different individuals may annotate the same image using different keywords, which make it difficult to create a suitable classification and annotate images with the exact keywords. To overcome all these reasons, content-based image retrieval (CBIR) is the area which is used for extracting images. The technique gray-level co-occurrence matrix (GLCM) is discussed and analyzed for retrieval of image. It considers the various features such as color histogram, texture, and edge density. In this paper, we mainly concentrate on texture feature for accurate and effective content-based image retrieval system.

Keywords CBIR · GLCM

P. Chandana (✉) · A. Gauthami Latha
Department of Computer Science Engineering, MES, Bhogapuram, India
e-mail: p.chandana1982@gmail.com

A. Gauthami Latha
e-mail: gauthamilatha@gmail.com

P. Srinivas Rao
Department of Computer Science Engineering, AU, Visakhapatnam, India
e-mail: peri.srinivasarao@yahoo.com

C.H. Satyanarayana
Department of CSE, JNTU, Kakinada, India
e-mail: chsaytanarayana@yahoo.com

Y. Srinivas
Department of IT, GITAM University, Visakhapatnam, India
e-mail: ysrinivasit@rediffmail.com

© Springer Nature Singapore Pte Ltd. 2017
S. Patnaik and F. Popentiu-Vladicescu (eds.), *Recent Developments in Intelligent Computing, Communication and Devices*, Advances in Intelligent Systems and Computing 555, DOI 10.1007/978-981-10-3779-5_4

1 Introduction

Definition

CBIR [1] signifies as query by image content (QBIC) [2] and content-based visual information retrieval (CBVIR). CBIR retrieves images based on the visual features such as color, texture, and shape. In image datasets, general methods of image indexing are difficult, insufficient, and time-consuming. In CBIR, each and every image that is stored in the database does feature extraction and matches with the query image. A CBIR system considers color [3, 4], texture, shape, and spatial locations as the low-level features of images in the datasets. When an input image or sketch is passed as input, the system retrieves similar images. This method is close to human perception of visual data and also reduces the requirement of describing the content in words. In CBIR, the word "content" describes about the context that states about the features such as colors, shapes, and textures. we can examine image content with the help of keywords. CBIR maintains the extracted features for both the dataset images and query images. In CBIR, every image in the dataset extracts its features and compares with the query image.

CBIR Process

Initialization process is the first step where it collects the images and considers all the images, and it determines the size of the image. Then, the feature extraction such as color [5] and location information is carried, and then, the extracted information is stored. Based on the information stored, the indexes are created and stored in the index table. The image database is processed off-line in order to save query-processing time [6, 7]. Feature extraction component is available for querying and retrieval only for the images that have been processed. In query process, user gives sample query image, and this query image goes through initialization, feature extraction, and indexing process [8, 9]. Then, the indexes are

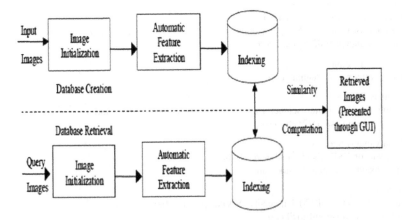

Fig. 1 Content-based image retrieval (CBIR)

compared with indexes in the index table. Images with similar index are displayed using graphical user interface (GUI). Figure 1 shows the components and the interaction carried in the proposed CBIR system.

2 Proposed System

Architecture of Feature Extraction Using GLCM
The architecture consists of two phases:

Phase 1   Preprocessing and
Phase 2   Feature extraction phase

Phase 1 processes the input data, and phase 2 extracts the features and combines both software and hardware to calculate GLCM features [10–12]. The GLCM feature vectors are calculated in hardware, and software supports hardware by performing additional computations. This is represented as in Fig. 2.

2.1 Preprocessing Phase

The preprocessing phase passes the conversion of the image into an array that is suitable for processing by the feature extraction phase. Each element $a = [a0, a1, a2, a4]$ of A corresponds to each pixel, and it is formed by five integers. $[a0]$ is the gray level of the corresponding pixel, and $[a1, a2, a3, a4]$ are the gray-level intensities of first neighbors in four directions. The resulting near quantization of the image intensities leads to 16-bit representation for each element.

Fig. 2 Architecture

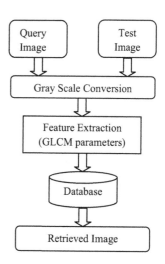

2.2 *Feature Extraction Phase*

The feature extraction [13] phase does the feature extraction based on the texture features. The GLCM [11] extracts the features of the image based on energy, contrast, entropy, correlation, etc. The extracted features are stored in the database, and the results are facilitated to the image search users through queries.

3 Methodology

Texture Feature Extraction based on GLCM

The texture features of an image are considered as the statistical properties for generating co-occurrence matrixes. The color image is converted to grayscale image, and then, we obtain image co-occurrence matrix [14, 15]. The content of an image is described using five properties [16, 17] such as contrast, energy, entropy, correlation, and local stationary. The properties are calculated by considering all the four directions, i.e., (i) horizontal (0°), (ii) vertical (90°), (iii) diagonal: (a) bottom left to top right (−45°) and (b) top left to bottom right (−135°). These are denoted as P_0, P_{45}, P_{90}, P_{270} and P_{135}, respectively.

The equations for identifying texture measures can be computed from gray-level co-occurrence matrices as shown below:

$$\text{Mean} = \sum_i \sum_j P(i,j) * i \tag{1}$$

$$\text{Variance} = \sum_i \sum_j P(i,j) * (i - \mu^2) \tag{2}$$

$$\text{Homogeneity} = \sum_i \sum_j \frac{P(i,j)}{1 + |i - j|} \tag{3}$$

$$\text{Contrast} = \sum_i \sum_j P(i,j) * (i - j)^2 \tag{4}$$

$$\text{Entropy} = \sum_i \sum_j -P(i,j) * \log_e P(i,j) \tag{5}$$

$$\text{Angular Second Moment} = \sum_i \sum_j P(i,j)^2 \tag{6}$$

$$\text{Correlation} = \sum_i \sum_j (i - \mu_x) * (j - \mu_y) * P(i,j) \tag{7}$$

$$\text{Dissimilarity} = \sum_i \sum_j \frac{1}{1 + (i-j)^2} * P(i,j) \tag{8}$$

$$\mu = \frac{\sum P(i,j)}{n} \tag{9}$$

$$\mu_x = \frac{\sum_j P(i,j)}{n}, \quad \mu_y = \frac{\sum_i P(i,j)}{n} \tag{10}$$

where $p(i, j)$ and n being the number of elements.

4 Similarity Feature Extraction

The similarity measure considered for comparison of images is Euclidean distance
In our paper, we concentrate on Euclidean distance, which applies the concept of Pythagoras' theorem for calculating the distance $d(x, y)$. The procedure for Euclidean distance [18] is given as follows:

$$d(x, y) = \sum_{t=1}^{n} |xi - yi|$$

The minimum distance value, i.e., range between 0 and 1 signifies an exact match with the query image. The distances in each dimension are squared before summation, and this gives much importance to find the dissimilarity. Therefore, the normalization the feature modules are necessary before finding the distance between the two images.

Other distance measure namely, KL Divergence, which is also referred as Kullback–Leibler distance (KL—distance) can be applied, which acts as a natural distance function from an "exact" probability distribution, p, to "required" probability distribution q. It can also be interpreted as expected message length per data obtained due to a wrong target (objective) distribution compared with respect to original distribution. For discrete data distribution, in which probabilities are defined as $p = \{p1, p2, p3..., pn\}$ and $q = \{q1, q2, q3..., qn\}$, the KL divergence is defined as

$$KL(p, q) = \sum_i p_i . \log_2(p_i/q_i).$$

For continuous distribution, the integral replaces the sum value.

$$KL(p, q) = 0$$
$$KL(p, q) \geq 0.$$

5   Algorithm for GLCM

Step1   Quantize the image data.
Step2   Create the GLCM [19].
Step3   Calculate the selected feature.
Step4   The sample s in the resulting virtual variable is replaced by the value of this calculated feature.

Stepwise description of GLCM:

1. Perform quantization on image data.
   The image is sampled and treated as a single pixel, and intensity is considered as a value for that pixel.
2. GLCM [14, 19, 20] is generated as follows:

$$P(i,j) = \sum_{x=1}^{N} \sum_{y=1}^{N} \begin{cases} 1, & \text{if } I(x,y) = i \text{ and } I(x+\Delta x, y+\Delta y) = j \\ 0, & \text{otherwise.} \end{cases}$$

It is a square matrix of size $N \times N$ where N is the **number of gray levels** specified during step 1. The offset Δx, Δy specifies distance between pixel of interest and its neighbor pixel.
The generation of matrix is done as follows:

   a. s is the sample that needs to be considered for calculation.
   b. W represents the sample set supporting sample s which is generated depending on the **window size**.
   c. By only considering the samples of set W, each element i, j of the GLCM is represented as the count that two samples of pixel intensities i and j occur in a window. The sum of all i, j in GLCM will be the number of times the specified spatial relationship occurs in W.
   d. Calculate symmetric GLCM matrix.

      i. Calculate transpose of GLCM.
      ii. Add the copy of GLCM to itself.

   e. Normalize the values of GLCM obtained by dividing each i, j with the sum of all elements in the matrix with respect to W.

3. Calculate the **selected feature**.
   This calculation uses the values in the GLCM as follows—energy, contrast, homogeneity, entropy, correlation, etc.
4. The sample s in resulting variable is replaced by the value of the calculated feature.

6 Results and Discussion

The query images are taken as input from the user, and after undergoing the preprocessing phase, it is converted to gray images as shown in Figs. 3 and 4. Figure 3 displays the details of number of files loaded, and in Fig. 4, it shows the features extracted such as angular second moment, contrast, correlation, IDM, entropy, and finally the sum of GLCM.

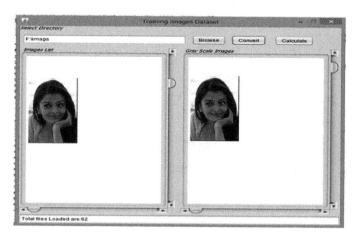

Fig. 3 Color image converted to gray image and gives number of files loaded (Color figure online)

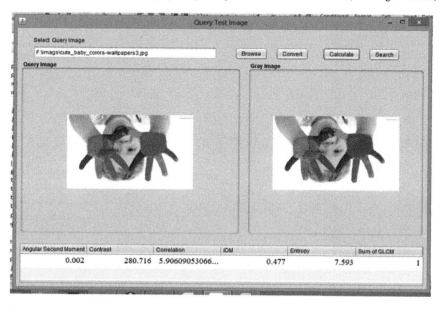

Fig. 4 Calculation of contrast, correlation, IDM, entropy, and sum of GLCM for color input image (Color figure online)

The features extracted from the feature extraction phase are stored in the database, and the result will be retrieved from the collection of image dataset based on the input query using distance and similarity measures. The results are shown below (Fig. 5).

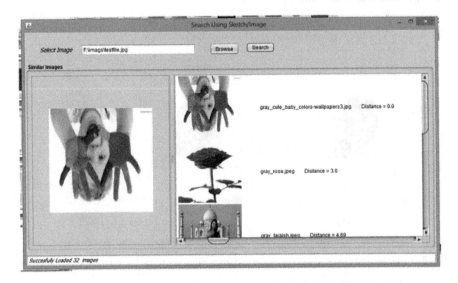

Fig. 5 Result is displayed with the distance matched and number of files loaded

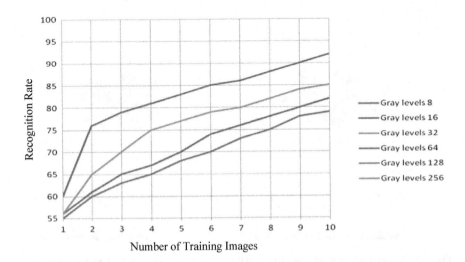

Fig. 6 Comparison using gray levels of training dataset

Table 1 Comparison of various methods

#Images	GLCM	LDA	PCA
8	98.67	97.33	94.5
16	87.54	90.54	90.65
32	83.34	85.64	84.32
64	80.56	78.67	75.47
128	79.7	75.89	73.98
256	78.65	77.9	76.54

7 Comparative Analysis

The below graph shows the recognition rate of images with respect to the number of images present in the training dataset by considering its gray levels as 8, 16, 32, etc (Fig. 6).

The following Table 1 shows the performance ratio of an image retrieval process based on the input query selected using different methods.

8 Conclusions

In this paper, we have proposed a new mechanism for CBIR systems, which is mainly based on two phases. The first phase is to convert the query image and the test image to gray-level images. The second phase is based on the feature extraction which includes texture features. These two processes will be undergone at both the side of training and testing images of the proposed system. At last, the similarity-based matching will be done where the distance of each observation is computed using the Euclidean distance measure and KL divergence approach. Finally, the normalized results were obtained by the proposed methodology. The experimental results show that the proposed technique of CBIR using GLCM retrieves the exactly matched images with the distances calculated. The distance ranges between 0 and 1 are the exactly matched images, whereas the distance range more than 1 is mismatched images.

References

1. I. Felci Rajam and S. Valli, "A Survey on Content Based Image Retrieval", Life Science Journal 2013, ISSN: 1097-8135
2. Yogita Mistry, Dr.D.T. Ingole, "Survey on Content Based Image Retrieval Systems", International Journal of Innovative Research in Computer and Communication Engineering, Vol. 1, Issue 8, October 2013, ISSN (Print): 2320-9798

3. Jianlin Zhang; Wensheng Zou "Content-Based Image Retrieval using color and edge direction features", Advanced Computer Control (ICACC), 2010 2nd International Conference on (Volume: 5), 27–29 March 2010
4. S.Meenachi Sundaresan, Dr. K.G.Srinivasagan, "Design of Image Retrieval Efficacy System Based on CBIR" International Journal of Advanced Research in Computer Science and Software Engineering, Volume 3, Issue 4, April 2013 ISSN: 2277 128X
5. Bansal M., Gurpreet K., and Maninder K, "International Content-Based Image Retrieval Based on Color", Journal of Computer Science and Technology, vol. 3, no. 1, pp. 295–297, 2012
6. Herráez J. and Ferri J, "Combining Similarity Measures in Content-Based Image Retrieval", Pattern Recognition Letters, vol. 29, no. 16, pp. 2174–2181, 2008
7. Kekre B., Mishra D., and Kariwala A., "A Survey of CBIR Techniques and Semantics", International Journal of Engineering Science and Technology, vol. 3, no. 5, pp. 4510–4517, 2011
8. Jayaprabha P. and Somasundaram M, "Content Based Image Retrieval Methods using Graphical Image Retrieval Algorithm", Computer Science and Application, vol. 1, no. 1. pp. 9–14, 2012
9. Manimala S. and Hemachandran H, "Content Based Image Retrieval using Color and Texture", Signal and Image Processing: An International Journal, vol. 3, no. 1, pp. 39–57, 2012
10. Biswajit pathak, debajyoti barooah, "Texture analysis based on the Gray-level co-occurrence Matrix considering possible Orientations", International Journal of Advanced Research in Electrical, Electronics and Instrumentation Engineering, Vol. 2, Issue 9, September 2013, ISSN: 2320 – 3765
11. Miroslav Benco, Robert Hudec, Patrik Kamencay, Martina Zachariasov, "An Advanced Approach to Extraction of Colour Texture Features Based on GLCM", Int J Adv Robot Syst, 2014, **11**:104 | doi:10.5772/58692
12. P. Mohanaiah, P. Sathyanarayana, L. Guru Kumar, "Image Texture Feature Extraction Using GLCM Approach", International Journal of Scientific and Research Publications, Volume 3, Issue 5, May 2013 1 ISSN 2250-3153
13. Manoharan Subramanian and Sathappan Sathappan, "An Efficient Content Based Image Retrieval using Advanced Filter Approaches", The International Arab Journal of Information Technology, Vol. 12, No. 3, May 2015, pgno: 229–236
14. M.Harsha vardhan, CS S.Visweswara Rao, "GLCM Architecture For Image Extraction", International Journal of Advanced Research in Electronics and Communication Engineering (IJARECE) Volume 3, Issue 1, January 2014, pg 74–82
15. Mryka Hall-Byer, "GLCM Texture: A Tutorial v. 2.7 April 2004", Department of Geography, University of Calgary, Canada
16. D. Sreenivasa Rao, N. Prameela, "Comparative Study on Content Based Image Retrieval Based on Color, Texture (GLCM & CCM) Features", International Journal of Science and Research (IJSR) ISSN (Online): 2319-7064, Volume 4 Issue 2, February 2015
17. Metty Mustikasari, Sarifuddin M, "Texture Based Image Retrieval Using GLCM and Image Sub-Block", International Journal of Advanced Research in Computer Science and Software Engineering, Volume 5, Issue 3, March 2015 ISSN: 2277 128x
18. Dr. Meenakshi Sharma, Anjali Batra, "An Efficient Content Based Image Retrieval System", IOSR Journal of Computer Engineering (IOSR-JCE) e-ISSN: 2278–0661, p- ISSN: 2278-8727 Volume 16, Issue 3, Ver. VI (May–Jun. 2014), PP 95–104
19. Sapthagiri.k, Manickam.L, "An Efficient Image Retrieval Based on Color, Texture (GLCM & CCM) features, and Genetic-Algorithm", International Journal Of Merging Technology And Advanced Research In Computing Issn: 2320–1363
20. GLCM Approach and textures representation http://support.echoview.com/WebHelp/Reference/Algorithms/Operators/GLCM_textures/
21. Amanbir Sandhu, Aarti Kochhar, "Content Based Image Retrieval using Texture, Color and Shape for Image Analysis", Council for Innovative Research International Journal of Computers & Technology, Volume 3, No. 1, AUG, 2012, ISSN: 2277-3061

Emotion Recognition System Based on Facial Expressions Using SVM

Ielaf Osaamah Abdul-Majjed

Abstract In recent years, there has been a growing interest in improving all aspects of interactions between human and computers particularly in the area of human expression recognition. In this work, we design a robust system by combining various techniques from computer vision and pattern recognition. Our system is divided into four modules started with image preprocessing, followed by feature extraction including Prewitt. Subsequently, the feature selection module has been performed as a third step using the sequential forward selection (SFS). Finally, support vector machine (SVM) is used as a classifier. Training and testing have been done on the Radboud Faces Database. The goal of the system was attaining the highest possible classification rate for the seven facial expressions (neutral, happy, sad, surprise, anger, disgust, and fear). 75.8% recognition accuracy was achieved on training dataset.

Keywords Facial expressions · SVM · SFS · Emotion recognition

1 Introduction

Different facial expressions reflected the human's inner feelings such as happiness or sadness; this emotion was used in computer vision system as an input to recognize these facial expressions. Most facial expression systems (FES) typically recognize the most widely used set which consists of six recognizable basic expression categories: happy, sad, surprise, anger, disgust, and fear (Fig. 1).

In this work, we develop a new system that includes four modules. First module illustrates image preprocessing in order to make the images more comparable. Secondly, feature extraction is executed in order to extract feature expression from an image before feeding the normalized image into the next step. Moreover, in

I.O. Abdul-Majjed (✉)
Department of Computer Science, College of Computer Science and Mathematics,
University of Mosul, Mosul, Iraq
e-mail: ielaf76@yahoo.com

© Springer Nature Singapore Pte Ltd. 2017
S. Patnaik and F. Popentiu-Vladicescu (eds.), *Recent Developments in Intelligent Computing, Communication and Devices*, Advances in Intelligent Systems and Computing 555, DOI 10.1007/978-981-10-3779-5_5

Fig. 1 Basic facial expression categories: happy, sad, surprise, anger, disgust, and fear [1]

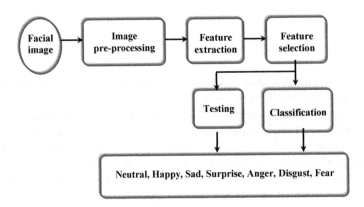

Fig. 2 Block diagram of FER system

order to reduce feature dimensions before classification, the feature selection module has been performed as a third step. Finally, the image has to be classified as SVM. Subsequently, the image has to be tested to classify it as one of the face expressions (neutral, happy, sad, surprise, anger, disgust, or fear) (Fig. 2).

2 Related Works

Among many researches that have been achieved in computer vision in order to develop techniques that provide facial expression recognition system.

A study was conducted in Massachusetts Institute of Technology by Skelley [2] to train a man–machine interface to recognize facial expressions, by using optical flow, texture value, and combined feature to extract the feature from Cohn-Kanade and HID database and then classified face expression by using null space principal component analysis (NSPCA) algorithm and support vector machines. 94.2% successful classification rate was achieved for the groups of subjects, and a 92% average for individuals [2].

A new method of facial expression recognition based on the SPE plus SVM by Ying et al. [3] used stochastic proximity embedding (SPE) for data dimension reduction and support vector machine (SVM) for expression classification. The proposed algorithm is applied to Japanese female facial expression (JAFFE) database for FER, better performance is obtained compared with some traditional algorithms, such as PCA and LDA [3].

Quadratic discriminant classifier (QDC), linear discriminant classifier (LDA), support vector classifier (SVC), and Naïve Bayes (NB) were introduced in 2007 by Wang and Yin [4] to classify Cohn-Kanade and MMI after topographic context (TC) expression descriptors.

Kotsis et al. [5] studied the effect of occluding 6 prototypic facial expressions on recognizing facial expressions. They applied three approaches for feature extraction: Gabor feature, by geometric displacement vectors extracted using Candide track and DNMF algorithm. After that they implemented multiclass support vector machine (SVM) and multilayer perceptron (MLP) on Cohn-Kanade and JAFFE databases. They indicated the recognition rate when using JAFFE: 88.1% with Gabor and 85.2% with DNMF, and when using the Cohn-Kanade: 91.6% with Gabor, 86.7% with DNMF, and 91.4% with SVM [5].

3 Emotion Recognition

Every emotion recognition system must perform a few steps before classifying the expression into a particular emotion. This works in four steps to recognize the emotion:

3.1 Image Preprocessing

In order to make the images more compared with reduced size, normalization has to be performed based on the alignment result by introducing two methods: clipping the images, and then performing histogram equalization to eliminate the variations in illumination and skin colors (Fig. 3).

3.2 Feature Extraction

Prewitt filter used to extract features expression from an image before feeding the normalized image into the next step (Fig. 4).

(a)   **(b)**

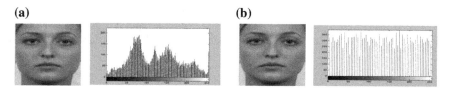

Fig. 3 Histogram equalization **a** clipping image and its histogram before histogram equalization and **b** the same image after histogram equalization and its histogram

Fig. 4 Prewitt edge detection
operator

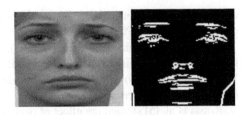

3.3 Feature Selection

In order to reduce feature dimensions before classification, feature selection has
been performed by using the sequential forward selection (SFS) method [6].

3.4 Expression Recognition

The final step of this system SVM was used to classify and tested the expressions
the input facial expressions into one of the seven types of ex-pression (Neutral,
Happy, Sad, Surprise, Anger, Disgust, Fear) [7–9].

3.5 Experiments Results

When applying the SVM method on the database, results achieved highest average
of classification rates equal to (75.8). Database resulted 100% accuracy was recognized in disgust, fear, happy, and natural expressions. The anger recognized as
easiest confused expression with recognition accuracy of 80%, except sad and
surprise were not recognized (Fig. 5).

Complete confusion matrix (Table 1) indicated that disgust, fear, happy, and
natural expressions were recognized with very high accuracy (100%). On the other
hand, anger expression recognized 80%.

Fig. 5 Complete confusion
matrix for selection
expression classification

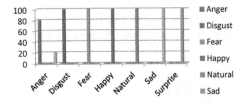

Table 1 Complete confusion matrix for selection expression classification

Expression	Anger	Disgust	Fear	Happy	Natural	Sad	Surprise
Anger	80	0	0	0	0	0	20
Disgust	0	100	0	0	0	0	0
Fear	0	0	100	0	0	0	0
Happy	0	0	0	100	0	0	0
Natural	0	0	0	0	100	0	0
Sad	0	0	0	0	0	0	100
Surprise	0	0	0	0	0	100	0

HMNN method used for classification after FCBF, SBS, and hybrid method

4 Conclusion

A new emotion recognition system based on the facial expression using SVM as a classifier was approached and tested on the Radboud Faces Database. The result of the suggested system was satisfactory; it achieved the highest average of classification rates equal to 75.8%. Four out of seven got very high accuracy (100%).

References

1. Langner, O., Dotsch, R., Bijlstra, G., Wigboldus, D.H.J., Hawk, S.T., & van Knippenberg, A. Presentation and validation of the Radboud Faces Database. Cognition & Emotion, 24(8), 1377–1388, 2010.
2. J. P. Skelley, "Experiments in Expression Recognition," MSc Thesis, Massachusetts institute of technology, 2005.
3. Zilu Ying, Mingwei Huang, Zhen Wang, and Zhewei Wang, "A New Method of Facial Expression Recognition Based on SPE Plus SVM," Springer-Verlag Berlin Heidelberg, Part II, CCIS 135, 2011, pp. 399–404.
4. M. Pantic, and I. Patras, "Dynamics of Facial Expression: Recognition of Facial Action and their Temporal Segment From Face Profile Image sequences," IEEE Trans. System, Man, and Cybernetic, Part B, Vol. 36, No. 2, 2006, pp. 433–449.
5. I. Kotsia, I. Buciu, and I. Pitas, "An Analysis of Facial Expression Recognition under Partial Facial Image Occlusion," Image and Vision Computing, Vol. 26, No. 7, 2008, pp. 1052–1067.
6. SF Pratama, AK Muda, YH Choo, NA Muda. "Computationally Inexpensive Sequential Forward Floating Selection for Acquiring Significant Features for Authorship," International Journal on New Computer Architectures and Their Applications (IJNCAA) 1(3): 581–598 The Society of Digital Information and Wireless Communications, 2011 (ISSN: 2220-9085).
7. V. N. Vapnik. "The Nature of Statistical Learning Theory". Springer, New York, 2nd edition, 2000.
8. S. V. N. Vishwanathan and M. N. Murty. "SSVM: A simple SVM algorithm", in Proceedings of IJCNN, IEEE Press, 2002.
9. Cortes C, Vapnik V. "Support-vector networks," Mach Learn, 1995, 20(3):273–297.

An AES–CHAOS-Based Hybrid Approach to Encrypt Multiple Images

Shelza Suri and Ritu Vijay

Abstract In today's world, one thing is for sure that technologies have advanced. Internet is being used as the primary mode of transfer of images, data or any other sort of textual information from one end to another. Thereby, it is very necessary to protect such mode of transfers from any sort of leakage of information to the outsiders which is something not intended by both the sender and the receiver. Hence, in order to protect such transferring of information, a hybrid approach to encrypt multiple images is being proposed. In such technique, fast chaotic encryption algorithm and the AES encryption algorithm are used together in order to create a hybrid approach to encrypt multiple images which are to be sent to the receiving end. On the receiver side, concepts of Cramer's rule are used in order to decrypt the images and make the algorithm to work in a more dynamic fashion, respectively.

Keywords Chaotic · AES · Cramer's rule · Watermarking

1 Introduction

With time, Internet is growing at a very fast pace. It can be definitely stated that in the communication world, internet is the primary mode of transfer of textual information or images from one end to another used by the human beings. However, it can be argued that nevertheless, technologies have made the lives of human beings easier, but these also bring with them certain drawbacks which are required to be tackled. During the transfer of information over the internet, certain security mechanisms should be used in order to protect the users' privacy that is there should be no kind of information leakage to any outsider. Such kind of security mechanism is encryption.

S. Suri (✉) · R. Vijay
Department of Electronics, Banasthali Vidyapith, Rajasthan, India
e-mail: Shelza_ecn@yahoo.com

R. Vijay
e-mail: rituvijay1975@yahoo.co.in

© Springer Nature Singapore Pte Ltd. 2017
S. Patnaik and F. Popentiu-Vladicescu (eds.), *Recent Developments in Intelligent Computing, Communication and Devices*, Advances in Intelligent Systems and Computing 555, DOI 10.1007/978-981-10-3779-5_6

Encryption can be defined as a process that converts plaintext into ciphertext and the reversal of such process that is the conversion of ciphertext into plaintext is known as decryption. Based on the encryption key, there are two different types of encryption which are public key encryption method where the key is sent along with the image and private key encryption method where the key is sent in an encrypted manner.

Image data do have special features such redundancy and correlation among pixels which do make encryption techniques tough to apply on them. However, there are various encryption techniques that can be used to encrypt data and sent to the receiver in a protected manner.

1.1 Fast Chaotic Algorithm

Firstly, chaos is derived from the Greek which means unpredictability and is defined as a study of nonlinear dynamic system. There is a general approach towards creating of chaos-based cipher designs which consists of the following 4 steps: (1) choose a chaotic map, (2) introduce the parameters that will be involved in the encryption process, (3) convert the values of the image into discrete values and (4) perform key scheduling.

In such algorithm, we involve the production of two logistic maps where the first logistic map is used to generate numbers from 1 to 24, and the second logistic map is modified constantly from the numbers generated by the first logistic map.

1.2 AES Image Encryption

There are three types of AES encryption that is AES-128, AES-192 and AES-256. All the modes in AES image encryption are done in 10, 12 or 14 rounds which depends on the block and the type of key being used. In our proposed algorithm, AES-128 is used. A larger key size results in more security and is faster as compared to the other AES encrypting algorithms.

It will be using four operation blocks which would operate on an array of bytes. The four stages are as follows: Substitute Bytes, Shift Row, Mix Columns and Add Round Key.

1.3 Cramer's Rule

In linear algebra, Cramer's rule is known as an explicit formula to find the solution of a linear system of equations with as many equations as the unknowns. It expresses the solutions in term of determinants which is evident from the figure shown below. The Cramer's rule will be very handy in decomposing the combined images so that these can be easily decrypted, and the receiver can get the images that were intended by the sender to be sent.

$$\text{Weight} = \begin{matrix} w_{11} & w_{12} \\ w_{21} & w_{22} \end{matrix} \tag{1}$$

Images are linearly combined using following equation:

$$I_1[i,j] = w_{11} * I'[i,j] + w_{12} * I''[i,j] \tag{2}$$

$$I_2[i,j] = w_{21} * I'[i,j] + w_{22} * I''[i,j] \tag{3}$$

The above equations would create an image that is a combination of two images that are multiplied by the values of the weight matrix. Images are decomposed using following equations:

$$I'[i,j] = \frac{\begin{vmatrix} w_{11} & I_1[i,j] \\ w_{21} & I_2[i,j] \end{vmatrix}}{\begin{vmatrix} w_{11} & w_{12} \\ w_{21} & w_{22} \end{vmatrix}} \tag{4}$$

$$I''[i,j] = \frac{\begin{vmatrix} I_1[i,j] & w_{12} \\ I_2[i,j] & w_{22} \end{vmatrix}}{\begin{vmatrix} w_{11} & w_{12} \\ w_{21} & w_{22} \end{vmatrix}} \tag{5}$$

The above equation resembles the working of Cramer's rule. Each pixel value of the image is obtained using the determinant of the weight matrix whose values of one column is interchanged with the values of the linearly combined. This value is further divided by the determinant of the original weight matrix. Same are performed for each pixel value of the image.

2 Literature Review

Chaos theory dates back to the 1970s. Researchers have found properties of chaos as a promising alternative for cryptographic algorithms. Its properties include sensitivity to initial conditions, ergodicity and quasi randomness. Avasare and Kelkar [1] performed two stage image encryption using two stages, that is transformation performed by implementing Chirkov mapping and encryption by discretizing the output of the first stage. Ismail et al. [2] introduced a new chaotic algorithm that is composed of two logistic map and an external key for image encryption of size 104 bits. Very recently, Chen et al. [3] proposed a symmetric image encryption in which a 2D chaotic map is generalized to 3D for designing a real-time secure image encryption scheme.

The Advanced Encryption Standard as described by Manoj and Harihar [4] was earlier known as Rijndael algorithm and uses symmetric key to protect sensitive information.

3   Proposed Algorithm

In the proposed algorithm, multiple images are transmitted simultaneously which are encrypted in a hybrid manner. Firstly, the images are taken and are encrypted using two different types of encryption methods, i.e. Image 1 would be encrypted by chaotic-based image encryption, while Image 2 would be encrypted by AES encryption algorithm. These two separate images would then be combined in order to create a final hybrid encrypted image that would be impossible to crack as there are two encryption algorithms applied on the image that is being sent (Figs. 1 and 2).

Once the image is received at the receiver side, the images are first decomposed using Cramer's rule in order to obtain the images that were encrypted by the algorithms listed above. After obtaining separate images, they are decrypted, respectively, according to the encryption technique applied on the image.

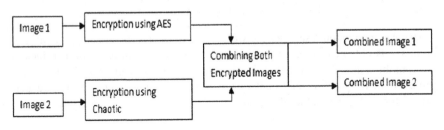

Fig. 1 Encryption process and combining images using Cramer's rule

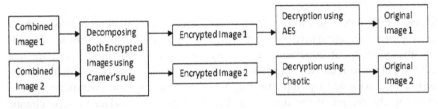

Fig. 2 Decomposing encrypted images using Cramer's rule and decryption process

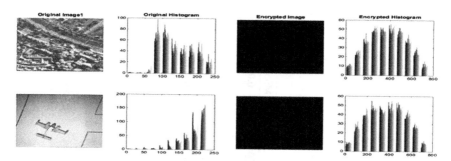

Fig. 3 Encryption of images with their histograms

4 Results

The proposed algorithm has successfully been applied on multiple images that is it is able to send multiple images simultaneously from the sender to the receiver in a more secure and efficient manner. Brief overview of the working of system is shown in Fig. 3. The following parameters have been used to measure the quality:

(1) Correlation Coefficient: The correlation coefficient refers to how different the encrypted image is from the original image. A correlation coefficient value of −1 refers to the negative of an image while a value of 1 refers to the image itself, and a value of 0 will refer to a complete different image. The correlation coefficients of different images are shown in Table 1.

(2) Mean Square Error: Mean square error is another form of quality measurement. It measures how different the reconstructed image is from the original image.

$$\text{MSE} = \frac{1}{n} \sum_{i=1}^{n} (Y_i' - Y_i) \tag{6}$$

where Y' is original image and Y is encrypted image.

Graphical user interface of the designed system using the proposed algorithm is illustrated in Fig. 4. The MATLAB version r2014b has been used for implementing the algorithm. As depicted by the figure, upload button is used to upload the images intended to be encrypted. The encrypt and decrypt buttons are used to initiate the encryption process and the decryption process, respectively. The save button is used to the save the images that have been encrypted or decrypted into our system. One has to enter the chaotic as well as the AES encryption key before clicking on the encrypt button.

Table 1 Results of different images

Algorithm	Parameter	Image name					
		Aerial	Airplane	Chemical plant	Clock	Resolution chart	Moon surface
AES	Correlation coefficient	−0.0012	0.0059	−0.0015	7.7258e+03	−0.0530	0.0012
	MSE	7.7258e+03	1.0882e+04	7.7991e+03	1.2141e+04	2.4081e+04	6.2262e+03
Chaotic	Correlation coefficient	0.0012	0.0113	0.0016	7.2478e−04	0.0107	−0.0047
	MSE	6.7663e+03	1.0903e+04	6.9151e+03	1.2439e+04	2.3396e+04	4.9749e+03
Proposed	Correlation coefficient	−0.0034	0.0100	−0.0094	9.1697e−004	9.1697e−004	−0.0060
	MSE	8.7646e+004	6.3883e+004	1.0643e+004	6.9049e+004	5.4453e+004	8.7492e+004

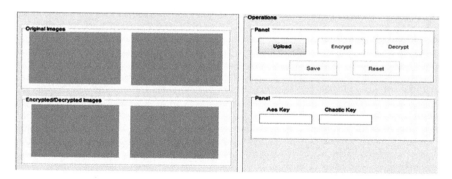

Fig. 4 Graphic user interface of implementation

5 Conclusion

The implemented work proposes multiple image transmission methods using different encryption methods simultaneously which are encrypted in a hybrid manner. It uses two different types of encryption methods considering merits of the both techniques. Hence, work can be extended by using some other techniques also.

References

1. Avasare, M. G., Kelkar, V. V.: Image encryption using chaos theory. In: Communication, Information & Computing Technology (ICCICT), International Conference on (pp. 1–6). IEEE (2015).
2. Ismail, I. A., Amin, M., Diab, H.: A digital image encryption algorithm based a composition of two chaotic logistic maps. IJ Network Security, 11(1), 1–10(2010).
3. Chen, G., Mao, Y., Chui, C. K.: A symmetric image encryption scheme based on 3D chaotic cat maps. Chaos, Solitons & Fractals, 21(3), 749–761(2004).
4. Manoj, B., Harihar, M. N.: Image encryption and decryption using AES. International Journal of Engineering and Advanced Technology (IJEAT), 1(5), 290–294(2012).

Automatic Text Summarization of Video Lectures Using Subtitles

Shruti Garg

Abstract Text summarization can be defined as a process of reducing a text document using computer program in order to generate a summary of original document that consists of most important things covered in that. An example of summarization technology is search engines such as Google. This paper orients for analyzing and producing text summary of video lectures by harnessing the subtitles file provided along with the lectures. Extractive text summarization method has been adopted to produce the summaries from the source subtitles. This would help user in deciding whether a particular lecture is relevant to them or not, thereby saving their time and aiding them in quick decision making. Experiments were conducted on various subtitle files belonging to different lectures, and it has been found that extractive text summarization reduces the content of original subtitle file up to sixty percent by tf-idf approach.

Keywords Summarization · tf-idf · Information retrieval · Subtitles

1 Introduction

Today's rapid pacing era of Internet, everyone wants to save time and get more information in lesser time. The field of text summarization helps people in such cases, and it prepares the summary of documents that can help very much in this regard. Summarization is the technique of producing most relevant information in the least possible number of words. Given a document, a text summarization system attempts to retrieve the best possible summary via an extractive methodology or an abstractive methodology which is more human readable. One of the mostly used application areas of text summarization is search engines such as Google and online news summarization to show only the relevant information to the person.

S. Garg (✉)
BIT, Mesra, Ranchi 835215, India
e-mail: gshruti_garg@yahoo.com

© Springer Nature Singapore Pte Ltd. 2017
S. Patnaik and F. Popentiu-Vladicescu (eds.), *Recent Developments in Intelligent Computing, Communication and Devices*, Advances in Intelligent Systems and Computing 555, DOI 10.1007/978-981-10-3779-5_7

The whole process of summarization can be divided into three basic steps/stages: (1) interpretation of source text into a suitable text format; (2) transformation of the text representation into a summary representation; and (3) finally, generation of summary text from the representation.

Text summarization systems are broadly classified into two categories—extractive and abstractive techniques. Extractive text summarization approach deals with presenting the sentences of original document into a compact format, and only those sentences are included into summary which have some relevance, whereas the abstractive approach works on rewriting the original text into a compressed format which is easily human readable.

Rene [1] and Luhn [2] find that the important terms in a document are selected by their term frequency and the position of that term affects the resultant summary. All the sentences are given a score which is an aggregate score of all the terms appearing in the sentence. This forms the basis of term frequency (tf) approach which was later used by many systems to statistically determine the importance of a sentence that is to be selected in the final abstract.

MEAD [3]: University of Michigan had developed the software MEAD in 2001. In their work, the extractive summaries of single and multi-document can be produced. The summaries were generated using centroid-based feature. Two more features were used by them, i.e., positioning and overlapping with the first sentence. Lastly, the linear combination of all three determines which sentences are most suitable to include in the final summary.

WebInEssence [4]: Again, University of Michigan had developed this software in 2001. It is used in more applications than a summarization system. This can be used as a search engine to summarize multiple-related Web pages that provide more contextual and summary information that help users to explore retrieval results better efficiently.

NetSum [5]: Microsoft Research Department developed NetSum in 2007 works for single-document summarization. This system is fully automated and produces single-document extracts of newswire articles using artificial neural nets. Artificial neural network is a machine learning approach which has two phases—training and testing. The trained set was labeled in such a way that the labels can identify the best sentences. After that, a set of features were extracted from every sentence in the train and test sets. After that, training is done using the same train cases. The distribution of features for the best sentences was learned by the system from these training sets, and the final output of the system is a ranked list of sentences for each document. The sentences were ranked using RankNet algorithm.

The remaining text of this paper is divided into two sections: In Sect. 2, the proposed system is explained, and in Sect. 3, the conclusion is provided.

2 Proposed System

In the proposed system of text summarization, the summarization of video lectures has been done, which are provided by many online resources such as NPTEL. The proposed approach exploits the subtitle files provided along the video lectures, performs the extractive text summarization on them, and produces a summary. It is a single-document-based summarization approach.

2.1 Proposed Architecture

In this system, author had used several steps where, first, the source file is gathered from the Internet, various data preprocessing techniques are used to clear the data, and then, several features are filtered out.

2.1.1 Data Preprocessing

- Removal of digits and frame numbers,
- Special character removal,
- Removal of newline characters,
- Stop words removal, and
- Lowercasing of the document.

2.1.2 Feature Selection for Summarization

For the generation of an effective summary, features used in the system are as follows:

- Content word feature
  Content words are usually called keywords are generally nouns in sentences, and these keywords have higher chances of being selected in the final summary.
- Sentence location feature
  This is based on the generalized fact that sentences present in the starting of a document and in the last paragraph of the document are of greater importance and have a higher tendency to be included in the final output.

- Sentence length feature

  The sentences which are very short in length usually do not have any relevant information and hence are not included in the final summary. Also, the sentences which are very long may not be containing any useful work and hence are not included in the summary.

- Cue-phrase feature

  The sentences containing phrases such as "in this lecture" and "most importantly," are of much importance and are included in the summary.

- Sentence-to-sentence cohesion

  In this step, the similarity between s and each other sentence s′ was calculated, and then, these similarity values were added for obtaining the first value of this feature for s. The same process is repeated for all sentences.

The example file taken is shown below:

```
00:00:51,699 --> 00:00:58,699
Today we start the lecture on the course Artificial
Intelligence. This course will be delivered
```

The result of data preprocessing and feature selection is shown below:

```
rtificial intelligence.
```

2.1.3   Processing After Feature Selection for Summarization

Once the above steps are completed, the tf-idf approach is used for the final summary generation.

- Term frequency (TF)

  The weighting of a term is done by considering that occurrence of that word in the document. For example if a term occurs more number of times in a document can have more reflect to the contents of document rather than a term that occurs less number of times. Thus, the weight of term is more if it has more relevance in document or having more frequency. The weight of term frequency is defined in terms of term frequency:

$$W_{t,d} = \begin{cases} 1 + \log_{10} tf_{t,d} & \text{if } tf_{t,d} > 0 \\ 0, & \text{otherwise} \end{cases}$$

Frequency of each term is given as:

```
{'breadth': 2, 'code': 4, 'partial': 1, 'consider': 7, 'represent': 3, 'global':
1, 'people': 3, 'leads': 3, 'concept': 1, 'children': 8, 'row': 6, 'whose': 5,
'graph': 3, 'deterministic': 1, 'to': 164, 'finally': 2, 'program': 3, 'nodes':
34, 'alternating': 2, 'applied': 2, 'very': 10, 'rise': 1, 'choice': 5, 'salespe
rson': 1, 'decide': 2, 'pply': 2, 'difference': 2, 'entire': 8, 'level': 4, 'pos
itions': 13, 'list': 5, 'solution': 10, 'try': 2, 'quick': 1, 'neighborhood': 1,
```

The TF scores are given as:

```
Computing the TF scores operation Successfull!!!
{'breadth': 1.6931471805599454, 'code': 2.386294361119891, 'partial': 1.0, 'cons
ider': 2.9459101490553135, 'represent': 2.09861228866811, 'global': 1.0, 'people
': 1.0, 'leads': 2.09861228866811, 'concept': 2.09861228866811, 'children': 1.0,
'row': 3.0794415416798357, 'whose': 2.791759469228055, 'graph': 2.6094379124341
005, 'deterministic': 2.09861228866811, 'to': 1.0, 'finally': 6.099866427824199,
'program': 1.6931471805599454, 'nodes': 2.09861228866811, 'alternating': 4.5263
```

- Inverse Document Frequency (IDF)

The term IDF was proposed by Salton for advancing information retrieval (IR) systems. If we count only the term frequency the problem arises that some terms which are irrelevant also get counted which does not give efficient summary. For example, the stop words generally have a high TF, but they are useless for discriminating the relevant documents since they tend to appear in most of the documents. IDF is defined as follows:

$$\text{IDF}(t) = \log_{10}(\text{Total no of documents} / \text{No of documents with term } t \text{ in it})$$

The idf scores are computed by the above formula, and the figure shows the values of idf scores of each term uniquely. Idf also calculates the relevancy of the term in a document that can also be considered as the importance of the term. The Lesser the occurrence of a term, the higher the importance will be.

```
The IDF Scores are :
{'consider': 0.12493873660829996, 'course': 0.6020599913279624, 'questions': 0.
0, 'previous': 0.3010299956639812, 'heuristic': 0.12493873660829996, 'situation
s': 0.6020599913279624, 'to': 0.3010299956639812, 'de': 0.6020599913279624, 'dc
': 0.0, 'cf': 0.6020599913279624, 'competing': 0.6020599913279624, 'cd': 0.1249
3873660829996, 'discuss': 0.3010299956639812, 'loss': 0.0, 'these': 0.0, 'where
': 0.6020599913279624, 'intelligence': 0.12493873660829996, 'cost': 0.124938736
```

S. Garg

The tf-idf approach provides a statistical measure for finding the importance of a given word in the document or among several documents. The importance of a word increases in proportional to the number of times a word appears in the document but is offset by the frequency of the word in the corpus. The tf-idf weighting scheme is popularly used by search engines for scoring and ranking of relevance of document given by a user as query. In the present work, tf-idf has been used within a single document instead of multi-document where it is usually applicable; however, in this, it has been applied on the collection of sentences and each of these has been treated as a document for correct results and evaluation purposes. The final tf-idf scores are computed by simply multiplying the tf scores of each term to its idf score.

```
The TF-IDF Scores are as follows:
{'code': 0.3010299956639812, 'move': 0.3010299956639812, 'judiciously': 0.301029
9956639812, 'leads': 0.0, 'previous': 0.0, 'to': 0.3010299956639812, 'finally':
0.0, 'then': 0.24987747321659992, 'return': 0.0, 'very': 0.0, 'familiar': 0.0, '
game': 0.0, 'successor': 0.0, 'loss': 0.0, 'this': 0.24987747321659992, 'slide':
0.3010299956639812, 'force': 0.0, 'computed': 0.0, 'play': 0.0, 'zero': 0.0, 'e
stimated': 0.0, 'best': 0.24987747321659992, 'plays': 0.3010299956639812, 'minim
ax': 0.24987747321659992, 'score': 0.0, 'squares': 0.24987747321659992, 'we': 0.
3010299956639812, 'backed': 0.3010299956639812, 'search': 0.3010299956639812, '1
```

- Normalization of tf-idf scores for sentence selection

Once the final tf-idf scores have been received for each unique term, now it is the time for the selection of sentences as sentences are formed by chunks of words/terms. So to find out the tf-idf weight of each sentence, the system adds up the weights of all the terms present in the sentence. This process is done for all the sentences present in the document.

However, all the sentences are not taken into the consideration of summary on the basis of their tf-idf scores alone. So, there is a need of a mechanism or methodology to normalize the score because there are always chances that a sentence having larger number of terms is always going to pass the threshold and get selected in the summary. In order to avoid this, a normalization approach has been used where the weight of each sentence is divided by the number of terms present in the sentence.

Norm(tf-Idf) = Sum of tf-idf score of all terms present in sentence/no of terms present in sentence.

```
Total Number of Sentences present in Document 1: 80
Total weight of all the sentences present: 8.27285245309
Average weight of each sentence: 0.103410655664
Number of sentences included in the Partial Summary: 39


Total Number of Sentences present in Document 2: 80
Total weight of all the sentences present: 3.94536279546
Average weight of each sentence: 0.0493170349433
Number of sentences included in the Partial Summary: 34


Total Number of Sentences present in Document 3: 80
Total weight of all the sentences present: 4.54310061643
Average weight of each sentence: 0.0567887577054
Number of sentences included in the Partial Summary: 33


Total Number of Sentences present in Document 4: 80
Total weight of all the sentences present: 4.88962221419
Average weight of each sentence: 0.0611202776774
Number of sentences included in the Partial Summary: 37


FINAL ANALYSIS

Number of sentences in original text : 320
Number of sentences in summary text : 143
Summary Percentage (No. of Sentences) : 44.6875
Summary Percentage (No. of Words) : 48.7263
```

Thus, this ensures that too long sentences do not get an unfair advantage in the selection criteria and also some relevant sentences having fewer number of terms do not get rejected because their high tf-idf scores would make sure that they are selected in the final summary.

For the selection of threshold value, the system takes the arithmetic mean weight of the whole document; all those sentences whose normalized score is higher than the average weight are finally selected in the summary; and the result is the collection of these sentences extracted from the source text.

$$\text{A sentence is selected if}: \text{tf-idf}(sent) \geq \text{Avg}(\text{tf - idf})$$

3   Conclusion

This work aims to provide a condensed and meaningful summary to the video lectures provided by different educational institutions or commercial Web sites (e.g., NPTEL, Coursera, and Udemy). The proposed system produces a summary which is nearly 44–45% of the original text source.

In terms of the number of words' reduction from the original source to that in the generated summary, the system successfully reduces the number of words in proportion to the summary generated, and also, it shows that the removal of stop words does not have an effect on the generated summary.

References

1. Arnulfo, R., Herandez, G., Ledeneva, Y.,: Word Sequence Models for Single Text Summarization, Second International Conferences on Advances in Computer-Human Interactions, IEEE, pp. 44–48, (2009).
2. Luhn, H. P.,: The Automatic Creation of Literature Abstracts. In Inderjeet Mani and Mark Marbury, editors, Advances in Automatic Text Summarization. MIT Press, (1999).
3. Lloret, E., Palomar, M.,: Text summarization in progress: a literature review, ACM journal of Artificial Intelligence Review, pp. 1–41, vol. 37 (1), (January 2012).
4. Radev, R., Blair-goldensohn, S., Zhang, Z.,: Experiments in Single and Multi-Document Summarization using MEAD. In First Document Under-standing Conference, New Orleans, LA, (2001).
5. Radev, D., Weiguo, F., Zhang, Z., Web in essence: A personalized web-based multi-document summarization and recommendation system. In NAACL Workshop on automatic Summarization, Pittsburg, (2001).

Classification of EMG Signals Using ANFIS for the Detection of Neuromuscular Disorders

Sakuntala Mahapatra, Debasis Mohanta, Prasant Kumar Mohanty and Santanu Kumar Nayak

Abstract Electromyography is used as a diagnostic tool for detecting different neuromuscular diseases and it is also a research tool for studying kinesiology which is the study of human- and animal-body muscular movements. Electromyography techniques can be employed with the diagnosis of muscular nerve compression and expansion abnormalities and other problems of muscles and nervous systems. An electromyogram (EMG) signal detects the electrical potential activities generated by muscle cells. These cells are activated by electrochemical signals and neurological signals. It is so difficult for the neurophysiologist to distinguish the individual waveforms generated from the muscle. Thus, the classification and feature extraction of the EMG signal becomes highly necessary. The principle of independent component analysis (ICA), fast Fourier transform (FFT) and other methods is used as dimensionality reduction methods of different critical signals extracted from human body. These different existing techniques for analysis of EMG signals have several limitations such as lower recognition rate waveforms, sensitive to continuous training and poor accuracy. In this chapter, the EMG signals are trained using soft computing techniques like adaptive neuro-fuzzy inference system (ANFIS). ANFIS is the hybrid network where fuzzy logic principle is used in neural network. This proposed technique has different advantages for better training of the EMG signals using ANFIS network with a higher reliability and better accuracy.

S. Mahapatra (✉) · P.K. Mohanty
Department of ETC, Trident Academy of Technology, BPUT,
Bhubaneswar, Odisha 751020, India
e-mail: mahapatra.sakuntala@gmail.com

P.K. Mohanty
e-mail: prsnt.mohanty@gmail.com

D. Mohanta · S.K. Nayak
Department of Electronic Science, Berhampur University,
Berhampur, Odisha 760007, India
e-mail: mohantadebasis@gmail.com

S.K. Nayak
e-mail: sknayakbu@rediffmail.com

© Springer Nature Singapore Pte Ltd. 2017
S. Patnaik and F. Popentiu-Vladicescu (eds.), *Recent Developments in Intelligent Computing, Communication and Devices*, Advances in Intelligent Systems and Computing 555, DOI 10.1007/978-981-10-3779-5_8

Discrete wavelet transformation (DWT) method is used for feature extraction of the signal.

Keywords Feature extraction · Classification · Electromyogram (EMG) · Neuromuscular disorders · Adaptive neuro-fuzzy inference system (ANFIS) · Discrete wavelet transformation (DWT)

1 Introduction

EMG (electromyogram) is a biomedical signal, which is the collection of the electrical signals, having different amplitudes and frequencies with respect to the function of time of the particular organ of the human body [1, 2]. EMG signal represents the electrical behaviour of the muscular organs of the human skin. The expansion and relaxation position of the muscular organs are controlled by the nervous system and is related to the physiological and anatomical properties of the muscles. The signals acquired from the human skin are generated due to the bio-chemical signals flowing inside the human body. The extracted EMG signals are low-amplitude weak signals, which are most of the times covered by the noises while travelling through the muscular tissues [3]. Clinical diagnosis and biomedical application of the EMG signal are the major interests of application. For powerful and sensitive applications, processing of the signal is a great challenge. Till date, scientists and researchers have given interest to develop better algorithms and methodologies to improve the analysis technique of the EMG signal with a better accuracy. If we are able to study the signal in a better accuracy, then we can investigate the actual problem of EMG signals. So far, so many traditional systems of algorithms have been used for the analysis of the signals [4–6]. Development of artificial intelligence systems and signal processing systems with advance mathe-matical tools give a new dimension for the research fields.

Elimination of the noises, detection and classification of the EMG signals are major parts of the work. Adaptive filter is used to eliminate the noises from the EMG signals; discrete wavelet transformation (DWT) is used to decompose the EMG signals, and adaptive neuro-fuzzy inference system (ANFIS) is used for the classification of the EMG signals.

2 Proposed Model

Detection and analysis of EMG signal consist the following steps [7]:

(a) Preprocessing of the EMG signal,
(b) Processing or features extraction of EMG signal, and
(c) Classification of the features of EMG signal.

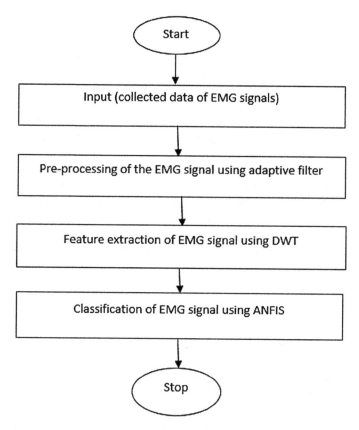

Fig. 1 Flow chart of proposed model

Figure 1 shows the flow chart of the whole algorithm of the proposed model.

2.1 Preprocessing of the EMG Signal

The collected EMG signals from the muscular organs are covered with different noises. These noises may appear basically due to two factors: one is from the biological factors and the other is from the instrumental sources. Muscular resistance, respiration, motion artefacts, muscular expansion and relaxation are the sources of noises related to the biological factors. Baseline drift noises, generated by electrodes of the instrument, and power line interference are the different sources of noises related to the instrumental factors. The EMG signals are manipulated due to these low-frequency noises may guide for the wrong diagnosis. Therefore, to reduce the noises, we have used the adaptive filter for EMG signal processing [3, 8].

2.2 Processing or Feature Extraction of EMG Signal

Wavelet transformation provides a better correlation between time and frequency domain. EMG signal, which is non-stationary in nature, can be easily analysed in both time and frequency domain by using DWT. Figure 2 shows the structure of wavelet packet tree (WPT), where $g(n)$ is the mother wavelet, $y(n)$ is the detail coefficient part and $x(n)$ is the approximation coefficient part of the WPT, after each level of decomposition. For the simplification of analysis and calculation, only low-frequency parts, i.e. approximation coefficient parts, are taken for the next level of decomposition. The maximum number of decomposition is dependent upon the size of data taken for calculation. $2N$ number of data samples can break down the signal into N discrete levels using DWT. Figure 2 shows three-level decomposition, but in our application, four-level decomposition is used. In the present model, we are employing sub-band analysis of the wavelet coefficient. Sub-band analysis is the method, which is a form of converting the coding that breaks the received signal into number of different frequency bands and analyses each one independently [9–11].

2.3 Classification of EMG Signal

ANFIS performs the human-like reasoning style of fuzzy systems. The use of fuzzy sets and a linguistic model in ANFIS system consist of a set of IF-THEN fuzzy rules [12–14]. The main strength of ANFIS is that they are universal approximators with the ability to implore explainable IF-THEN rules. The adaptive neuro-fuzzy system considered in this chapter is described in Fig. 3. A two-input first-order Sugeno fuzzy model with two rules has been described below; x and y are two inputs, and f is the output.

Fig. 2 Wavelet packet tree

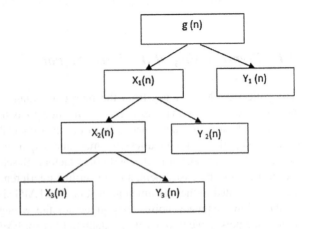

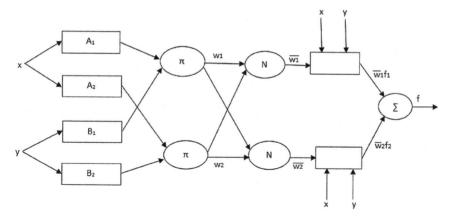

Fig. 3 ANFIS network

Rule – 1

If x is A_1 and y is B_1, then   (1)

$$f_1 = p_1x + q_1y + r_1$$

Rule – 2

If x is A_2 and y is B_2, then   (2)

$$f_2 = p_2x + q_2y + r_2$$

In the first layer of ANFIS network, all nodes are the adaptive nodes; x and y are the inputs, and A_i and B_i are the linguistic labels.

$$A_i, B_i, \quad \text{where } i = 1, 2.$$

Membership functions (pi and bell) for fuzzy sets are as follows:

$$\mu_{Ai}, \mu_{Bi} \quad \text{where } i = 1, 2.$$

Every node labelled with π in layer two is the fixed node whose output is the product of all the inputs. Each node represents the firing strength of the rule.

$$w_i = \mu_{Ai}(x) \cdot \mu_{Bi}(y), \quad i = 1, 2 \tag{3}$$

Every node in layer three, labelled by N, is the fixed node. The ith node calculates the ratio of the ith rule's firing strength to the sum of all rules' firing strength. These outputs are the normalized values.

$$\overline{W}_i = w_i/[w_1 + w_2], \; i = 1, 2 \tag{4}$$

Each node in layer four is an adaptive node with a node function.

$$\overline{W}_i f_i = \overline{W}_i (p_i x + q_i y + r_i) \tag{5}$$

Then, the single output node f is a fixed node, which computes the overall output as the summation of all the input signals.

$$f = \overline{w1} f_1 + \overline{w2} f_2 \tag{6}$$

3   Results and Analysis

The original signal is fed to DWT, and the four levels of decomposed output are shown in Fig. 4. Comparison plot between ANFIS input and output is shown in Fig. 5. The error minimized for both in ANFIS and neuro-fuzzy network is shown in Fig. 6.

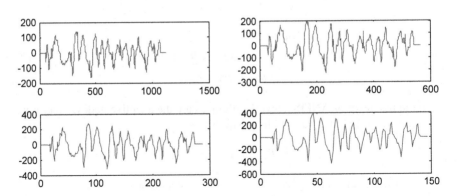

Fig. 4 Four-level decomposed output

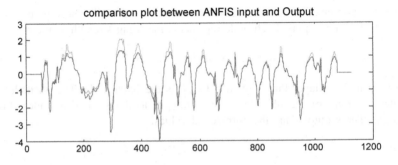

Fig. 5 Comparison plot between ANFIS input and output

Fig. 6 Training curve of
ANFIS and neuro-fuzzy

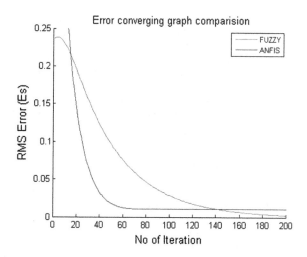

4 Conclusion

This chapter represents an ideal approach for the EMG signal analysis. The analysis and classification of both the signals are carried out using the ANFIS and neuro-fuzzy network. As shown in Fig. 6 (comparison of training curves between ANFIS and neuro-fuzzy network), ANFIS network provides a better error minimization as compared to the neuro-fuzzy network. Due to better error minimization and higher accuracy of the proposed method, ANFIS can be used as a real-time EMS signal classification system. Higher accuracy of the system makes it highly reliable and efficient. An ANFIS network has the advantage of combining the best features of fuzzy systems and neural networks in the biomedical signal classification.

References

1. Adnan Ahmed Ali; Sch. of Inf. Eng., Wuhan Univ. of Technol., Wuhan, China Albarahany; Liu quan... "EMG signals detection technique in voluntary muscle Movement" 738–742 (ISSDM), IEEE, 2012.
2. Jolanta Pauk, "Different techniques for EMG signal processing", Journal of Vibroengineering, 571–576, December 2008.
3. Sakuntala Mahapatra, Santanu k. Nayak, Samrat L. Sabat, "Neuro Fuzzy model for adaptive Filtering of oscillatory Signals", Elsevier Science, Measurement 30, 231–239, 2001.
4. Jeffrey Winslow; Adriana Martinez; Christine K. Thomas "Automatic Identification and Classification of Muscle Spasms in Long-Term EMG Recordings" IEEE Journal of Biomedical and Health Informatics Year: 2015, Volume: 19, Issue: 2, Pages: 464–470.
5. Direk Sueaseenak; Sunu Wibirama; Theerasak Chanwimalueang; Chuchart Pintavirooj; Manus Sangworas "Comparison Study of Muscular-Contraction Classification Between Independent Component Analysis and Artificial Neural Network" IEEE, ISCIT 2008. Pages: 468–472.

6. Y. Matsumura; Y. Mitsukura; M. Fukumi; N. Akamatsu; Y. Yamamoto; K. Nakaura "Recognition of EMG signal patterns by neural networks" IOCNIP'02, IEEE, Year: 2002, Volume: 2, Pages: 750–754.
7. M.B.I. Reaz, M.S. Hussain, F. Mohd-Yasin "Techniques of EMG signal Analysis: detection, processing, classification and applications" December 2006, Springer science, Volume 8, Issue 1, pp 11–35.
8. Tokunbo Ogunfunmi, Thomas Paul "Analysis of Convergence of Frequency-Domain LMS Adaptive Filter Implemented as a Multi-Stage Adaptive Filter" springer Science, 341–350, 2008.
9. Abdulhamit Subasi, Mustafa Yilmaz, Hasan Riza Ozcalik "Classification of EMG signals using wavelet neural network" Journal of Neuroscience Methods. Volume 156, Issues 1–2, 30 September 2006, Pages 360–367.
10. Zhaojie Ju; Gaoxiang Ouyang; Marzena Wilamowska-Korsak; Honghai Liu "Surface EMG Based Hand Manipulation Identification via Nonlinear Feature Extraction and Classification" Year: 2013, Volume: 13, Pages: 3302–3311, IEEE sensor Journal.
11. Zhu Xizhi "Study of Surface Electromyography Signal Based on Wavelet Transform and Radial Basis Function Neural Network" 2008 International Seminar on Future Bio Medical Information Engineering, IEEE, Pages: 160–163, 2008.
12. Jang, J.-S.R., "ANFIS: Adaptive-Network-Based Fuzzy Inference Systems," IEEE Transactions on Systems, Man, and Cybernetics, Vol. 23, No. 3, pp. 665–685, May 1993.
13. Jang, J.-S.R. and C.-T. Sun, Neuro-fuzzy modeling and control, Proceedings of the IEEE, March 1995.
14. Sakuntala Mahapatra, Raju Daniel, Deep Narayan Dey, Santanu Kumar Nayak, "Induction Motor Control Using PSO-ANFIS", Procedia Computer Science (Elsevier), Vol. 48, pp. 754–769, 2015.

Evaluating Similarity of Websites Using Genetic Algorithm for Web Design Reorganisation

Jyoti Chaudhary, Arvind K. Sharma and S.C. Jain

Abstract Web mining is a necessary and important process of data mining that automatically discovers the vital information from Web documents and Web services. Web mining is the method of retrieving the beneficial information from activities held over the World Wide Web. It can be classified into three different forms such as Web usage mining, Web content mining and Web structure mining. Websites are essential tools for the Web users to obtain necessary information such as education, entertainment, health and e-commerce from the Web. The purpose of this paper is to present a genetic algorithm to evaluate the similarity of two or more Web pages of the Websites. Besides it, the complexity of Website design will be computed by reorganizing the Web pages of the Website. The obtained results would definitely be increasing the effectiveness of the Websites.

Keywords Web mining · WSM · Genetic algorithm · PHP · Java

1 Introduction

Website is a collection of Web pages, and it is in a semi-structured location called Web. Many researches have been carried out for the improvement of the Web. Discovery and analysis of pattern are the techniques useful for Web personalization. Web marketing becomes more popular in the media to bring more visitors to the particular Website. Web mining uses new direction in the Web marketing. Web

J. Chaudhary (✉) · S.C. Jain
Department of CSE, Rajasthan Technical University, Kota, India
e-mail: chaudharyjyoti1990@gmail.com

S.C. Jain
e-mail: scjain1@yahoo.com

A.K. Sharma
Department of CSI, University of Kota, Kota, India
e-mail: drarvindkumarsharma@gmail.com

© Springer Nature Singapore Pte Ltd. 2017
S. Patnaik and F. Popentiu-Vladicescu (eds.), *Recent Developments in Intelligent Computing, Communication and Devices*, Advances in Intelligent Systems and Computing 555, DOI 10.1007/978-981-10-3779-5_9

mining has the potential to find out the target customers. Exploring the Weblog will not give a clear idea about the user of the Website. Nowadays, due to rapid growth of the World Wide Web, a wealth of data on many different subjects and topics are available online. Websites are playing an important role not only for companies but also for private individuals trying to find diverse information. Websites are good communication channels to provide useful information to the end-users. Many Websites are growing and indirectly lead to increase the complexity of Website design. To simplify the complexity of Website design, we have to understand the structure of website and their working. This task is achieved by using one of the techniques, called Web mining [1]. The objects of Web mining are vast, hetero-geneous and distributing documents. The logistic structure of Web is a graph structured by documents and hyperlinks, and the mining results may be on Web contents or Web structures. The activities such as Web designing, creating attractive Websites, are also a part of Web usage mining techniques. Because of its direct application in e-commerce, Web analytics, e-learning, information retrieval etc., Web mining has become one of the important areas in computer and information science.

The rest of this paper is structured as follows: Sect. 2 discusses Web mining background. Section 3 presents literature survey. Section 4 contains proposed methodology. In Sect. 5, experimental evaluation is shown while conclusion is given in the last section.

2 Web Mining—A Background

Web mining is a combination of data mining and WWW. Web mining is the data mining technique that automatically discovers and extracts the information from the Web documents [2]. The significant Web mining applications are Website design, Web search, search engines, information retrieval, network management, e-commerce, business and artificial intelligence, Web marketplaces and Web communities. Web mining has three classifications, namely Web content mining, Web structure mining and Web usage mining. Each classification is having its own algorithms and tools.

2.1 Web Mining Taxonomy

Web mining is classified into three categories such as Web usage mining, Web content mining and Web structure mining. The Web mining taxonomy is shown in Fig. 1 below.

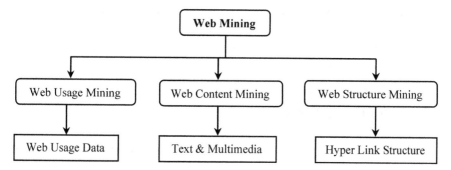

Fig. 1 Web mining taxonomy

Fig. 2 Web mining tasks

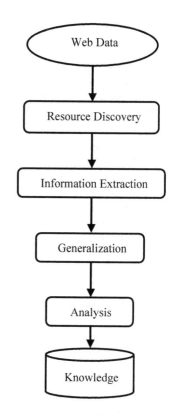

- **Web Usage Mining**—It is a method by which we can predict the behaviour of user. It helps to identify the browsing pattern using user behaviour. It uses a secondary data for this purpose [3].
- **Web Content Mining**—It is the method of obtaining beneficial knowledge from the contents of Web pages. Content data correspond to the collection of facts a Web page was designed to convey to the users [2].

- **Web Structure Mining**—It is the method describing the link structure of the Web pages in Web documents. The aim of Web structure mining is to develop structured synopsis about the Web pages or Websites [4].

2.2 Web Mining Tasks

The Web mining process can be achieved by using the following tasks which are shown in Fig. 2.

3 Literature Survey

Many researchers have contributed in the area of Web mining algorithms. The following section discusses the various works of several authors. Some works of the researchers are discussed here.

In Neelam Tayagi et al. [5] presented comparative study of three Web structuring algorithms page ranking algorithm, weighted page rank algorithm and hypertext-induced topic search algorithm. In M.B. Thulase et al. [6] proposed an algorithmic framework for reorganizing the Website using splay and heap tree structures. In Zaiping Tao [7] has discussed an efficient WTPM algorithm based on the pattern growth framework for resolving the issue of mining transaction pattern. In Shanthi et al. [8] presented a strategy for automatic Weblog information mining by Web page collection algorithm. In Ramanpreet Kaur et al. [9] proposed and implemented an enhanced version of AdaBoost algorithm in Web mining. In Mini Singh Ahuja et al. [10] presented a link analysis algorithm which has divided into two categories: algebraic and probabilistic method. In Shesh Narayan Mishra et al. [11] proposed a new concept based on topic-sensitive page rank and weighted page rank for Web page ranking. In Swapna Mallipeddi et al. [12] presented high-utility mining algorithm for preprocessed Web data. Sharma et al. [13] proposed a decision tree algorithm and existing C4.5 algorithm for comparative study and analysing the performance. In Guang Feng et al. [14] was proposed a novel algorithm aggregate rank and proved mathematically that could well approximate the probability of visiting a Website. In Alexandros Nanopoulos et al. [15] proposed a new data mining algorithm WMO for generalized Web prefetching. In Mohammed J. Zaki [16] was presented new algorithm SPADE for the discovery of sequential patterns. In Arun Kumar Singh et al. [17] discussed a novel page rank algorithm for Web mining based on user's interest.

Fig. 3 Work flow of
proposed genetic algorithm

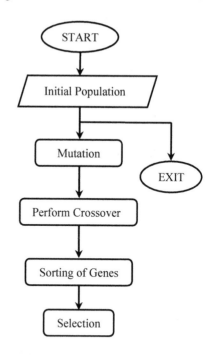

4   Proposed Methodology

Website evaluation determines needed modification in the contents of Website and link structure of Website. The technique for Website evaluation is to model user navigation pattern and compare them to site designer's expected patterns. In this section, an effective approach for genetic algorithm has been proposed and presented. Genetic algorithms are used to handle a population of possible solutions, and each solution is expressed through a chromosome which is an abstract representation. Genetic algorithm is used to solve optimization problems. In this work, a genetic algorithm is shown and it would be returning a chromosome for each Web application. These chromosomes are represented by a binary tree that keeps a set of distinct text from the Web application. The work flow diagram of our proposed genetic algorithm is shown in Fig. 3.

W1.php, w2.php, w3.php, w4.php and w5.php are the examples of Web pages which are presented below. We use in-order traversal to present the binary tree.

For web page w1.php:

```
<?php>
<html>
<title> My first paper </title>
<head>
<body>
```

Web application 1

```
<u>RTU1</u>
<i>RTU2</i>
</head>
</body>
</html>
<?>
```

With the help of given PHP codes, we create a novel binary tree of the Website for:

```
<?>
</html>
</body>
</head>
</i>RTU2<i>
</u>RTU1<u>
```

Web application 1

```
<body>
<head>
</title> My first paper <title>
<html>
<?php>
```

Similarly, the Web code for another Web page is as follows:

```
<?php>
<html>
<title> My second paper is </title>
<head>
<body>
```

Web application 2

```
<u>MTECH 1</u>
<i>MTECH 2</i>
</head>
</body>
</html>
<?>
```

In the same way, another similar binary tree is also created.

```
<?>
</html>
</body>
```

```
</head>
</i> MTECH 2 <i>
</u> MTECH 1 <u>
```

Web application 2

```
<body>
<head>
</title> My second paper is <title>
<html>
<?php>
```

The PHP code for this Web page is as follows:

```
<?php>
<html>
<title> My third research paper is </title>
<head>
<body>
```

Web application 3

```
<u>PHD 1</u>
<i>PHD 2</i>
</head>
</body>
</html>
<?>
```

In the same way, another similar binary tree is also created that is as follows:

```
<?>
</html>
</body>
</head>
</i>PHD2<i>
</u>PHD1<u>
```

Web application 3

```
<body>
<head>
</title> My third research paper is <title>
<html>
<?php>
```

Similarly, code for another Web page is as follows:

```
<?php>
<html>
```

```
<title> My fourth research paper is </title>
<head>
<body>
```

Web application 4

```
<u>MTECH 3</u>
<i>MTECH 4</i>
</head>
</body>
</html>
<?>
```

Similarly, we evaluate another binary tree for Web code w4.php

```
<?>
</html>
</body>
</head>
</i>MTECH 4<i>
</u>MTECH 3<u>
```

Web application 4

```
<body>
<head>
</title> My fourth research paper is <title>
<html>
<?php>
```

For Web page w5.php:

```
<?php>
<html>
<title> My fifth paper is </title>
<head>
<body>
```

Web application 5

```
<u>UCEE 1</u>
<i>UCEE 2</i>
</head>
</body>
</html>
<?>
```

Similarly, we find another binary tree for Web code w5.php

```
<?>
</html>
</body>
</head>
</i>UCEE 2<i>
</u>UCEE 1<u>
```

Web application 5

```
<body>
<head>
</title> My fifth paper is <title>
<html>
<?php>
```

5 Experimental Evaluation

In this section, we have proposed a genetic algorithm to evaluate the similarity for different Web pages by using PHP and HTML codes. The complete experiment is to be executed on Window 8 platform, Core i3 processor, 1.70 GHz with 4.00 GB RAM. This proposed algorithm would be implemented using PHP and Java language.

5.1 Proposed Genetic Algorithm

A genetic algorithm has been designed for creating a binary tree and for computing a longest common substring for the two Websites. Our algorithm contains two main steps. In the first step, we calculate the significant text for each Web page of the Website. In the second step, we compute the similarity value by using a specific technique. This algorithm containing genetic elements would be started with population of chromosome characterized by set of Web pages and a binary tree. We have a sequence, one for mutation and one for selection. We have a fitness function (performance function). We have created an array in which element 'a' gives the number of distinct text from chromosome 'a'. The best chromosome is one with the biggest number of distinct tags. We, therefore, create a binary tree for each Web page containing a set of significant tag.

The pseudocode of the proposed algorithm is as follows:
Binary Tree ()

```
for i = 1; Nr pages to do
select a page;
select a tag and add it into tree;
```

```
increase a no. of tags from tree;
end for
for i = 1, Nr binary tree do;
inorder traversal;
determine string;
end for
```

LCS ()

```
for i = 1; S1.length do
for j = 1; S2.length do;
if S1[i] = S2[j] then
increase length of LCS;
add leaf character used in another string;
else
retain character with the largest character to obtain long-
est LCS;
continue comparing;
end if
end for
end for
```

5.2 Computation of Similarity

In this section, we discuss an expression as a formula to compute the similarity value among the Web pages of the Website.

Let C be the common substring obtained from the Web pages.

So that, we consider C1 as a string obtained from the binary tree of the host Web page w1. Similarly, C2 for second Web page w2, C3 for third Web page w3, C4 for fourth Web page w4 and C5 for fifth Web page w5.

Let L be the length of common substring C.

Length of C1 is L1, for C2 is L2, for C3 is L3, for C4 is L5 and for C5 is L5. Similarity can be calculated between the Web pages as

$$Sw = \frac{T}{(L1 + L2 + L3 + L4 + L5)}$$

From this, we find a general formula to calculate the similarity value among the Web pages, which is as follows:

$$Sw = \frac{T}{\sum_i^n Li} \tag{1}$$

Sw   Similarity value
T   Total number of Web pages
Li   Length of common substring Ci

6   Conclusion

In this paper, a genetic algorithm has been proposed to find the similarity of Web pages for the two Websites. The presented algorithm comprises of two necessary steps. In the first step, the significant text is calculated, and in another step, the similarity value of Web pages is evaluated. We have also tried to identify exact common elements from two Web pages of the Websites by using genetic algorithm. The binary trees are used to keep the tags from the significant Web pages. The proposed algorithm has been implemented in PHP and Java language. As future work, we shall propose to implement this algorithm for different Web development languages such as ASP and JSP.

References

1. Arvind K. Sharma et al., "Enhancing the Performance of the Websites through Web Log Analysis and Improvement", International Journal of Computer Science and Technology, Vol. 3, Issue-4, 2012.
2. Dushyant Rathod, "A Review on Web mining", International Journal of Engineering Research and Technology, Vol. 1, No. 2 (April-2012), ESRSA Publications, 2012.
3. Md. Zahid Hasan et al., "Research Challenges in Web Data Mining", IJCST, Vol. 3, Issue-7, 2012.
4. Dhanashree S. Deshpande, "A Survey on Web Data Mining Applications", ETCSIT (2012).
5. Neelam Tyagi et al., "Comparative study of various page ranking algorithms in Web Structure Mining", IJITEE, Vol. 1, Issue-1, 2012.
6. M. B. Thulase et al., "An Algorithmic Framework for Re-Organizing the Website Using Splay and Heap Tree Structures".
7. Zaiping Tao, "A Fast Web Transaction Pattern Mining Algorithm", IJCSITS, Vol. 2, No. 2, 2012.
8. R. Shanthi and Dr. S.P. Rajagopalan, "An Efficient Web Mining Algorithm to Mine Web Log Information", IJIRCCE, Vol. 1, Issue-7, 2013.
9. Ramanpreet Kaur and Vinay Chopra, "Implementing Adaboost and Enhanced Adaboost Algorithm in Web Mining", IJARCCE, Vol. 4, Issue-7, 2015.
10. Mini Singh Ahuja and Sumit Chhabra, "A Review of Algebraic Link Analysis Algorithms", IJCAIT, Vol. 1, Issue-2, 2012.
11. Shesh Narayan Mishra et al., "An effective algorithm for web mining based on topic sensitive link analysis", IJARCSSE, Vol. 2, Issue-4, 2012.
12. Swapna Mallipeddi et al., "High Utility Mining Algorithm for Preprocessed Web Data", IJCTT, Vol. 3, Issue-3, 2012.
13. Pooja Sharma et al., "Implementation of Decision Tree Algorithm to Analysis the Performance", IJARCCE, Vol. 1, Issue-10, 2012.

14. Guang Feng et al., "Aggregate Rank: Bringing order to Web sites", Proceedings of the 29th annual International ACM SIGIR conference on Research and development in information retrieval, ACM, 2006.
15. Alexandros Nanopoulos et al., "A data mining algorithm for generalized web prefetching", IEEE Transactions on Knowledge and Data Engineering, Vol. 15, Issue-5, 2003.
16. Mohammed J. ZAKI, "SPADE: An efficient algorithm for mining frequent sequences", Machine Learning 42.1–2 (2001): 31–60.
17. Arun Kumar Singh and Niraj Singhal, "A Novel Page Rank Algorithm for Web Mining based on User's Interest", IJETAE, Vol. 2, Issue-9, 2012.

Fusion of Misuse Detection with Anomaly Detection Technique for Novel Hybrid Network Intrusion Detection System

Jamal Hussain and Samuel Lalmuanawma

Abstract Intrusion detection system (IDS) was designed to monitor the abnormal activity occurring in the computer network system. Many researchers concentrate their efforts on designing different techniques to build reliable IDS. However, individual technique such as misuse and anomaly techniques alone failed to provide the best possible detection rate. In this paper, we proposed a new hybrid IDS model with feature selection that integrates misuse detection technique and anomaly detection technique based on a decision rule structure. The key idea was to take the advantage of naïve Bayes (NB) feature selection, misuse detection technique based on decision tree (DT), and anomaly detection based on one-class support vector machine (OCSVM). First, misuse detection is built using single DT algorithm where the training data get decomposed into multiple subsets with the help of decision rules. Then, anomaly detection models are created for each decomposed subset based on multiple OCSVM. In the proposed model, NB and DT can find the best-selected features to ameliorate the detection accuracy by obtaining decision rules for known normal and attack anomalies. Then, the OCSVM can detect new attacks that result in an improvement in the detection accuracy of classification. The proposed new hybrid model was evaluated based on the NSL-KDD data sets, which is an upgraded version of KDD99 data set developed by DARPA. Simulation results demonstrate that the proposed hybrid model outperforms conventional models in terms of time complexity and detection rate with the much lower rate of false positives.

Keywords Hybrid IDS · Feature selection · Naïve Bayes classifier · Decision tree · One-class SVM

J. Hussain · S. Lalmuanawma (✉)
Mathematics and Computer Science Department, Mizoram University,
Tanhril, Aizawl, Mizoram 796004, India
e-mail: samuellalmuanawma@mzu.edu.in

J. Hussain
e-mail: jamal.mzu@gmail.com

© Springer Nature Singapore Pte Ltd. 2017
S. Patnaik and F. Popentiu-Vladicescu (eds.), *Recent Developments in Intelligent Computing, Communication and Devices*, Advances in Intelligent Systems and Computing 555, DOI 10.1007/978-981-10-3779-5_10

1   Introduction

An intrusion detection system (IDS) has been developed for the second line of defense in the security environment. The conventional prevention systems such as data encryption, user validation, and firewalls are implemented as the first line of defense for the computer security [1]. However, intruders know how to detour these defense tools; as a result, the second line of defense is required, which is constituted by tools such as IDS and antivirus software [2].

IDS aims to help filter out various potential unauthorized accesses to the computer network system. It gathers and analyzes information data from various sources within the computer network, triggers alarm to the system administrator, and blocks unauthorized access if an attack attempt is found.

IDS is categorized into two main techniques: misuse method and anomaly method. In misuse method, the observed behaviors were compared against the predefined attack pattern known as a signature, and an alarm is raised if a match is found. Misuse-based network intrusion detection system (NIDS) has been widely deployed in recent network communications security. However, due to its limitation over network packet overload, expensive computational power on signature matching, and massive false-negative rate, research has been carried out using anomaly detection technique. Anomaly method tried to detect the abnormal activity from the normal user behavior, and if the observed behavior deviates too much from the normal baseline profile, an alarm was triggered to the system administrator [3]. However, these detection systems still suffered from some limitations such as ineffective detection rate toward known attack and massive false-positive rate.

To resolve the disadvantages of these two conventional IDS techniques, a hybrid intrusion detection system (HIDS) combining both misuse and anomaly techniques has also been proposed by recent research [4, 5]. In a hybrid technique, both misuse and anomaly methods are combined in such a way to ameliorate the performance of detection accuracy along with the lower degree of false alarm. The performance of the detection depends on the combination method of these two conventional detection techniques. Most HIDS trains both anomaly and misuse detection techniques independently and then calculates the weighted average results of the detection technique [5]. These techniques obviously increase the detection rate but still have a high degree of false alarm. To overcome this situation, Kim et al. [4] proposed a new technique that integrates misuse and anomaly techniques but still has high time complexity due to the absence of feature selection.

The purpose of this research is to develop a new hybrid approach to overcome the limitation of both misuse and anomaly techniques for the NIDS. The proposed technique involves two steps that hierarchically incorporate misuse detection technique based on C4.5 decision tree (DT), and an anomaly detection technique using one-class SVM with feature selection using naïve Bayes to ameliorate the classification accuracy. The proposed model is also compared with other conventional techniques, and the results demonstrate that the proposed technique outperforms the conventional models.

2 Proposed Hybrid Intrusion Detection Methodology

This paper proposes a new hybrid novel NIDS with feature selection by fusing misuse detection technique with anomaly detection technique based on the decision rule structure using misuse technique that results in decomposed subsets of the original data sets. Then, anomaly detection technique based on the multiple one-class SVM classifications for each decomposed subset was designed to detect an outlier from the normal baseline profile. The key idea of the proposed hybrid detection technique is to combine the advantages of misuse detection technique well known for its low-level false-positive rate and anomaly detection technique that can detect the novel or unknown attack traffic. Figure 1 demonstrates the proposed model involving three stages: (i) feature preparation module, (ii) misuse analyzer module, and (iii) anomaly analyzer module. Details of these three modules are discussed in the following subsections.

Fig. 1 Proposed hybrid detection technique fusing misuse and anomaly techniques

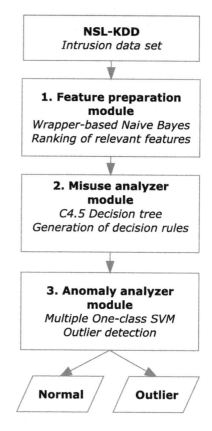

2.1   Feature Preparation Module

Feature selection or feature extraction is the main technique used by the recent research to improve the performance of a classification model. The main purpose of feature selection is to reduce the computational time of the classification technique and improve the classification performance by removing redundant and unrelated attributes to find a subset of features for the proposed approach [6], and this was accomplished by removing redundant or unrelated features out of its original set. Removal of important or relevant features might decrease the performance of the classification model regarding detection rate. However, some features might have a high degree of noise or might not have a contribution in any way. Removal of such features can significantly improve the detection accuracy and search speed of a classification model [7]. Generally, there are two major types of feature selection methods: filter method and wrapper method [8, 9].

In this module, feature selection based on naïve Bayes is used. The naïve Bayes classifier was used to identify the relevant features and rank them accordingly to create a subset of features for every normal and attack instances.

Naïve Bayes classifier is well known for its robustness against noise and missing values, high-speed training time, and simplicity in approach with clear semantics and interpretation [10]. In this paper, the NB classifier was used within a wrapper-based feature selection technique. The wrapper-based method uses the classifier to identify the subsets of relevant features by evaluating the subsets and correlations of each feature [11] within a greedy stepwise wrapper method. The proposed feature selection was evaluated using NSL-KDD intrusion data set [12], which is an enhanced version of KDD99 [13] intrusion data set created by DARPA. We use two-class classification strategies along with k-fold cross-validation (K-FCV) technique within the proposed feature selection module. K-FCV is one of the most commonly used methods where data set gets divided into k, in which k represents the number of folds or subsets, $k - 1$ subsets are used as a training set, and $k - (k - 1)$ subsets are used as testing sets. More specifically, each feature was analyzed regarding relevancy for each fold, and then, the total score results determine the most relevant features out of k-folds.

Classification approach of the naïve Bayes classifier is based on the probabilistic method. It typically relies on the assumption and assumes that variables are independent within each class or feature. More specifically, the presence of each particular class or feature is isolated in the absence or occurrence of some other feature. Given a vector of random variable data point, $X = (x_i, \ldots, x_n)$, denoting the observed attribute value, the naïve Bayes classification algorithm will predict that data point $\times$ test case belongs to the class c (i.e., normal or abnormal) random variable denoting the class of instances with the maximum posterior probability:

$$P(x|c) = \prod_{i=1}^{n} P(x_i \mid c) \tag{1}$$

2.2 Misuse Analyzer Module

In this module, a misuse detection model was designed based on one of the most frequently used classification algorithms called DT. DT utilizes a divide-and-conquer approach and recursively creates a DT based on the greedy algorithm [14, 15]. DT consists of the root node, branches, parent nodes, child nodes, and leaf nodes. It constructs a treelike structure in a series of Boolean formation 'yes' or 'no' until no more related branches can be derived. A node in a tree denotes data set attributes; every child node derives labeled branches in relation to the possibilities of attribute values from the corresponding node called parent node [4]. A branch connects either one or two nodes or a leaf, and each labeled leaf node represents the classification value. A classification for new data is obtained starting from the root node and moving down toward the branches until and unless a leaf node is found, a decision rule has been created to categorize the data point according to the value of features.

DT calculates and selects the maximum value of information gain (Eq. 2). To select those features having a maximum value of information gain, DT starts from the root node and then divides the data set into more subsets, until no more relevant branches can be derived or until all data in the current subset fit into an identical class. If $C_1 \ldots, C_n$ denotes classes and T is for DT representing a leaf that identifies class C_1, then the information gain ratio is calculated as follows [16]:

$$\text{Info Gain}(F) = \text{Info}(Y) - \text{Info}_x(G)$$

$$\text{Info}(Y) = - \sum_{i=1}^{k} \frac{\text{freq}(C_i, Y)}{|Y|} \times \log_2 \left(\frac{\text{freq}(C_i, Y)}{|Y|} \right) \tag{2}$$

$$\text{Info}_x(G) = \sum_{j=1}^{n} \frac{|G_j|}{|G|} \times \text{Info}(G_j)$$

Here, C_i denotes classes where i arts from 1 to k, k is the maximum number of classes, $|Y|$ represents total cases amount of train set, the standard quantity of information required to categorize the case in class Y is represented by $\text{Info}(Y)$, for partition G the $\text{Info}_x(G)$ denotes those features F having an applicable amount of information, total amount of cases integrated in C_i is represented by $\text{freq}(C_i, Y)$, n represents total quantity of outputs intended for F, G_j represent T subset in relation to jth term, and $|G_j|$ denote the total cases amount of the G_j subset.

In this work, a misuse analyzer module for the proposed hybrid model was designed based on DT, using the normal and attack data to train the model. DT divides input data into more decomposed regions based on the information gain. The decomposed subsets created by this module serve as input to the next-level anomaly-based classification technique. To get an optimal performance of the DT algorithm, it is required to set the important parameters such as confidence value, a number of folds for cross-validation, and minimum cases [17].

2.3   Anomaly Analyzer Module

In this module, an anomaly-based detection technique was designed based on the one-class SVM classification algorithm [18]. It is popularly known for its outlier detection ability for various applications. Classification algorithms that make use of only one class label, mainly normal class, is called one-class classifier. A one-class classifier needs to be trained before it can classify any data points.

Based on the normal profile decomposed structure from the misuse analyzer module, we trained multiple OCSVM classifiers to create a multiple normal baseline profiles, throughout the training phase each model locate the decision margin separation between the inlier and outlier instances. In the inspection phase, the decision function of each one-class model detects outlier connections that could be attacked. An outlier can be known or unknown attack, while the inliers are those normal activities.

Vapnik [19] first proposed the SVM model based on the idea of increasing dimensionality of the binary class samples so that they can be separable. The basic idea of SVM is to find a maximum hyperplane to separate binary samples from the same class inside it. The extension of SVM, OCSVM model, was proposed by Schölkopf et al. [20]. It was formulated to find a maximum hyperplane that separates a desired portion of the one-class training instances in feature space (F) from its origin. In the testing phase, the outliers of a testing instance are detected based on this hyperplane, to determine which class the instance belongs.

Let us consider a training data $x_1, \ldots, x_l \in X$, where X is the original space and the number of instances denoted by l. Let ϕ be a feature map $\chi \to F$ to locate a hyperplane that best separates training data pattern from the original space X, which transforms the instances nonlinearly to the feature space from its original so as to establish the best hyperplane in F. The one-class SVM is formulated as follows:

$$\min_{w,\xi,p} \frac{1}{2}\|w\|^2 + \frac{1}{vl}\sum_{i=1}^{l}\xi_i - p \tag{3}$$

subject to...

$$(w.\phi(x_i)) \geq p - \xi_i,$$
$$\xi_i \geq 0, \ i - 0, \ldots l$$

where p represents the distance of the origin from the hyperplane, w is the vector orthogonal to the hyperplane, and ξ_i is a slack variable of a vector $\xi_i, \ldots, \xi_l$. As mentioned in Perdisci et al. [21], there are some difficulties in the calculation of the feature space which is caused by the curse of dimensionality; then, simple kernel function $k(x, y) = (\phi(x). \phi(y))$, such as Gaussian $k(x, y) = e^{-\gamma\|x-y\|2}$, was utilized to compute the feature space. Each instance (n) was generally tested by a function

$f(n)$ which returns the decision results of which side of the hyperplane each encountered instance falls on in the feature space, formulated as follows:

$$f(n) = \text{sgn} \left(\sum_{i=1}^{l} \left(\alpha_i k x_{i(n-p)a} \right) \right) \tag{4}$$

If the $f(n)$ returns a positive value, it means the encountered instance belongs to the inlier feature space, but if there is a negative result, it means the encountered instance is an outlier. In this paper, the term 'inlier' indicates the normal activity, as this anomaly module is built based on only the normal profile from the decomposed subset of the misuse module, while outliers indicate those attacks which might harm the system, i.e., known or unknown attack.

3   Simulation Results

The proposed hybrid system is evaluated carefully based on the NSL-KDD intrusion data set [12]. The NSL-KDD data set is an enhanced version of KDD99 intrusion data set prepared at the MIT Lincoln Laboratories, USA, by the DARPA. The KDD99 data set contains a lot of redundant records, where 78 and 75% are duplicate on training and test data set which may direct machine learning algorithm toward unreasonable classification results. The NSL-KDD data set was proposed for resolving such problems and is found to be more efficient to have more realistic environment compared to the KDD99 intrusion data set. To evaluate the performance of the proposed hybrid system, Weka 3.7.11 [22] and LibSVM MATLAB [18] is used. The proposed hybrid system was evaluated as follows.

First, feature selection is done on the preprocessed data set from NSL-KDD data, and we organized the evaluation data by modifying the original data set KDDTrain +_20Percent with KDDTrain+ and KDDTest-21 with KDDTest+. The author transforms the ordinary multiclass data set into two-class data set so that they can be used for the evaluation of the proposed hybrid system. This modification is done to include unknown attack to the test data set; unknown attack means attacks traffic data that has neither been used for training nor been seen by the network before. Then, feature selection module selects those relevant features from the original 41 features of the NSL-KDD data set based on the wrapper feature selection with k-fold cross-validation technique. The wrapper-based method uses the naïve Bayes classifier to identify subsets of relevant features by evaluating the subsets and correlations of each feature within a greedy stepwise wrapper method. Table 1 illustrates the feature sets for the proposed hybrid system. The main idea of the feature selection in this paper is to reduce the computational time of the classification algorithm and improve the classification performance by removing redundant and unrelated attributes. Evaluation results (Table 3) demonstrate that the time complexity of the proposed misuse detection is reduced to 18.48 s compared to the

Table 1 Selected feature sets for proposed technique based on naïve Bayes feature selection

Sl. no.	Feature name	Data type	Description
1	duration	Continuous	Length of the connection
2	protocol_type	Nominal	Connection protocol
3	service	Nominal	Destination service
4	src_bytes	Continuous	Bytes sent from source to destination
5	su_attempted	Nominal	1 if 'su root' command attempted; 0 otherwise
6	count	Continuous	Number of connections to the same host as the current connection in the past two seconds
7	srv_count	Continuous	Number of connections to the same service as the current connection in the past two seconds (same service connections)
8	dst_host_srv_count	Continuous	% of connections having the same destination host and using the same service
9	dst_host_same_srv_rate	Continuous	% of connections having the same destination host and using the same service
10	dst_host_serror_rate	Continuous	% of connections to the current host that have an S0 error
11	dst_host_rerror_rate	Continuous	% of connections to the current host that have an RST error

conventional methods [4], achieving much more detection rate with an acceptable rate of false alarm.

Once the feature set is identified, they were taken into the misuse classification stage where the data get divided into more decomposed subsets. The basic idea of decomposing the normal data into multiple subsets is that the conventional hybrid anomaly model intended to profile the normal activity based on an outlier detection technique. However, there is multiple normal activities' profile based on the types of service, protocol, src_byte, etc., in a real environment. The anomaly analyzer based on OCSVM can be extremely responsive to the train set and may lead toward a high degree of false alarm rate. To resolve this situation, the normal data set gets decomposed into more appropriate subsets, and then, a single one-class SVM classification model is built for each subset. After applying the identified features to a C4.5 DT, the original data set get decomposed into multiple subsets based on decision rules created by the DT classification algorithm (Table 2) based on the information gained from each feature.

It is observed that the classification results of the proposed system outperform the conventional models [4] in terms of time complexity, detection rate, false-positive rate, and root mean square error (Table 3). To obtain the optimal performance of the DT algorithm, important parameters are carefully set until an optimal result is obtained. A compatible detection rate of 99.99% with only 0.1% false alarm rate with 0.0107 RMS errors was achieved after setting the minimum instance per leaf approximately to 1.0% and the confidence factor of 1%. It was also

Table 2 Decision rules obtained by proposed misuse detection technique based on C4.5 decision tree

Sl. no.	Decision rules	Type
1	src_bytes> 28 and src_bytes<= 333 and dst_host_srv_count<= 205 and service = domain_u	Normal
2	src_bytes<= 28 and dst_host_srv_count<= 89 and count <= 6 and src_bytes> 1 and count <= 1	Normal
3	src_bytes> 28 and src_bytes<= 333 and dst_host_srv_count<= 205 and service = ftp and dst_host_srv_count> 2	Normal
4	src_bytes> 28 and src_bytes<= 333 and dst_host_srv_count<= 205 and service = ftp_data and dst_host_same_srv_rate<= 0.94	Normal
5	src_bytes<= 28 and dst_host_srv_count<= 89 and count <= 6 and service = http and dst_host_serror_rate<= 0.5 and dst_host_same_srv_rate> 0.25	Normal
6	src_bytes> 28 and src_bytes<= 333 and dst_host_srv_count<= 205 and service = http	Normal
7	src_bytes<= 28 and dst_host_srv_count> 89 and src_bytes<= 0 and dst_host_serror_rate<= 0.7 and service = http	Normal
8	src_bytes> 28 and src_bytes<= 333 and dst_host_srv_count<= 205 and service = other and src_bytes<= 145 and dst_host_serror_rate<= 0	Normal
9	src_bytes> 28 and src_bytes<= 333 and dst_host_srv_count<= 205 and service = private and src_bytes<= 156 and src_bytes> 102	Normal
10	src_bytes > 333 and src_bytes > 334 and service = ecr_i and service = ftp_data and dst_host_same_srv_rate <= 0.98 and dst_host_serror_rate <= 0.01	Normal
11	src_bytes> 28 and src_bytes<= 333 and dst_host_srv_count<= 205 and service = telnet and dst_host_rerror_rate<= 0.27	Normal
12	src_bytes> 333 and src_bytes> 334 and service = ecr_i and service = ftp_data and dst_host_same_srv_rate<= 0.98 and dst_host_serror_rate<= 0.01	Normal
13	src_bytes > 28 and src_bytes <= 333 and dst_host_srv_count <= 205 and service = urp_i and dst_host_same_srv_rate <= 0.35	Normal
14	src_bytes<= 28 and dst_host_srv_count<= 89 and count > 6 and src_bytes<= 10	Attack
15	src_bytes> 333 and src_bytes> 334 and service = ecr_i	Attack
16	src_bytes<= 28 and dst_host_srv_count<= 89 and count <= 6 and service = http and dst_host_same_srv_rate<= 0.06	Attack
17	src_bytes> 333 and src_bytes<= 334 and service = ftp_data	Attack
18	src_bytes> 40494 and dst_host_same_srv_rate<= 0.98 and duration > 1398 and service = http	Attack
19	src_bytes<= 28 and dst_host_srv_count<= 89 and count <= 6 and service = eco_i	Attack
20	src_bytes> 334 and src_bytes<= 40494 and service = telnet and dst_host_same_srv_rate> 0.75 and su_attempted<= 0	Attack
21	src_bytes<= 28 and dst_host_srv_count<= 89 and count > 6 and src_bytes> 10 and protocol_type = udp	Attack

(continued)

Table 2 (continued)

Sl. no.	Decision rules	Type
22	src_bytes> 333 and src_bytes> 334 and service = ecr_i and service = ftp and duration <= 13	Attack
23	src_bytes<= 28 and dst_host_srv_count> 89 and src_bytes> 0 and service = eco_i	Attack
24	src_bytes<= 28 and dst_host_srv_count<= 89 and count <= 6 and service = ftp_data and src_bytes<= 4 and srv_count<= 6 and duration <= 2511 and dst_host_serror_rate> 0.51	Attack

Table 3 Misuse detection based on C4.5 DT

Methods	Conventional model [14]	Proposed model
Feature selection	NULL (41)	Wrapper-based naïve Bayes (11)
Time taken to build model (s)	23.22	4.74
TPR	99.75	99.99
FPR	0.3	0.10%
RMS error	0.0492	0.0107

found that the time complexity of the misuse model decreases up to 79.6% as compared to the conventional methods.

Once the decomposed structure is established, a multiple OCSVM classification algorithm model is built based on each normal activity. Conventional anomaly detection system builds a classification algorithm based on only the normal traffic information. The proposed hybrid model also followed this method; however, a decomposed structure of the normal training instances is proposed to improve the normal activity profiling performance of the anomaly module. Because the whole normal traffic has a range of normal associations, there is a problem on profiling those models accurately for an anomaly technique and can degrade the performance [23].

Each decomposed subset was tested on multiple one-class SVM with normal traffic and creates a decision function that describes the normal behavior that separates the inlier from the outlier. Each one-class SVM model was carefully evaluated based on the important parameter γ with variation from 0.001 to 1 and compared with the conventional models. Too narrow or too broad in the parameter γ may affect the detection problem on detecting unknown attack for OCSVM. In this experiment, it is observed that the best value for parameter γ is 0.01. An increase in parameter γ results in an elevation of detection performance of OCSVM model (Fig. 2). The detection performance was investigated based on the anomaly detail, and this was achieved by testing various kernels such as linear, polynomial, sigmoid, and Gaussian $k(x, y) = e^{-\gamma\|x-y\|^2}$. In Fig. 2, it is observed that Gaussian kernel outperforms other kernels in terms of detection accuracy of 99.98% along

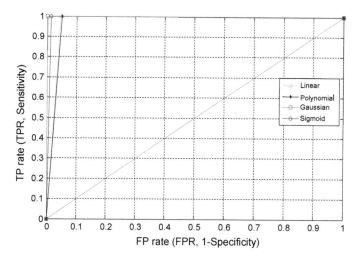

Fig. 2 ROC curve performance of various kernels in OCSVM

Fig. 3 Performance of various kernels (time complexity) in OCSVM

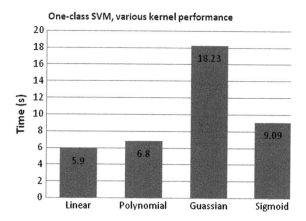

with much lower false-positive rate of 0.1%. However, it requires much time complexity compared to others (Fig. 3), but the main focus of these studies is to improve the ability of unknown attack detection rate along with an acceptable rate of false alarm and time complexity. As a result, training time and testing time are calculated based on Weka and MATLAB applications (Table 4). As expected, the time complexity of the proposed model is improved to 37.97 s (training time) and 6.71 s (testing time), which is shorter than the conventional models 280.99 s (training time) along with 19.17 s (testing time). So, we can conclude that the time complexity of the proposed model is improved up to 86% (training time) with 65% (testing time) compared to the conventional models. This was achieved due to the application of feature selection procedure and decomposition of the original data set into more decomposed subsets. Since each decomposed data structure is less

Table 4 Comparison of detection time between conventional model and the proposed new model

Decomposed subset @ decision rules	No. of training instances	Training time (s)	No. of testing instances	Testing time (s)	No. of support vectors
Subset 1	2507	1.39	499	0.06	251
Subset 2	10,681	26.41	9840	5.5	2136
Subset 3	261	0.02	1389	0.014	48
Subset 4	1399	0.42	834	0.057	126
Subset 5	685	0.17	7942	0.42	130
Subset 6	68	0.01	907	0.005	7
Subset 7	176	0.02	164	0.004	62
Subset 8	7564	8.8	434	0.18	2042
Subset 9	1001	0.41	390	0.29	230
Subset 10	209	0.03	269	0.005	75
Subset 11	519	0.11	339	0.013	36
Subset 12	34	0.01	49	0.0006	12
Subset 13	685	0.17	7942	0.42	199
Proposed model	25,789	37.97	30,998	6.71	5353
Conventional model [4]	25,789	280.99	30,998	19.17	7891

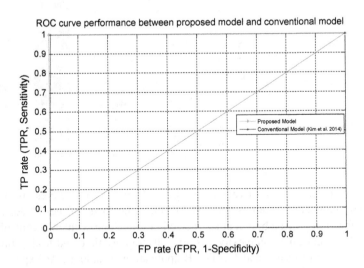

Fig. 4 ROC curve comparison of proposed model with the conventional models

complex as compared to the original data structure, a single one-class SVM model for the whole original data pattern can be more flexible than multiple OCSVM models for each decomposed subset.

ROC curve performance (Fig. 4) and comparison (Table 4) demonstrate that the proposed model outperforms the conventional models regarding detection rate of unknown attack and training and testing time complexity. 32% reduces the average number of encountered support vectors for the proposed model compared to the conventional models. Since the number of support vector affects the complexity of the testing computation, a lesser number of avg_SVs is suggested to improve the time complexity of the classification algorithm.

4 Conclusion

In this paper, we proposed a new hybrid NIDS, which integrates feature selection, misuse-based detection system, and anomaly-based detection technique. Our proposed model includes wrapper-based feature selection to improve the IDS regarding detection performance on the unknown attack and improve the training/testing time complexity of a classification algorithm. The key idea of our proposed hybrid model is to combine the advantages of misuse detection, well known for its low-level false-positive rate, and anomaly detection techniques, which can detect the novel traffic activity.

First, a naïve Bayes classifier based on the wrapper method is used to identify relevant features and rank them accordingly to create a subset of features, and this attempt focuses on the particular attribute that describes both attack and normal activities of intrusion data. These reductions of feature also filtered out those non-relevant features that associate with noisy data and also decrease the computational power. Then, a misuse-based detection model is designed based on C4.5 DT that decomposed the original data set into smaller decomposed subsets. Then, multiple anomaly-based detection model was created based on multiple OCSVM for each decomposed subset. The anomaly-based detection model used the normal activity to create the normal baseline profile; any deviation from the model is treated as outliers. Outliers could be known or unknown attack traffic. The experimental results demonstrate that the proposed system can improve the intrusion detection regarding novel attack detection and time complexity of IDS.

Finally, we have concluded that our proposed new hybrid IDS technique results in the improvement on both misuse and anomaly detection techniques. The time complexity of misuse and anomaly-based is reduced up to 86.5 and 65% with an overall accuracy of 85.1%. It is also encountered that the average number of support vectors is reduced up to 32% with a high detection rate of 99.98% along with an acceptable rate of 0.1% false alarm.

These evaluation results encourage us for further research on various hybrid IDS techniques, and exploration of various classification algorithms against real network traffic along with the effect of various noisy data over machine learning algorithm may be the focus of our future works.

References

1. Lazarevic, A., Ertoz, L., Kumar, V., Ozgur, A., Srivastava, J. (2003). A comparative study of anomaly detection schemes in network intrusion detection. In Proceedings of the 3rd SIAM Conference on Data Mining.
2. Lee, J. H., Sohn, S. G., Chang, B. H., Chung, T. M. (2009). PKG-VUL: Security vulnerability evaluation and patch framework for package-based systems. ETRI Journal, 31(5), 554–564.
3. Beauquier, J., Hu, Y. (2008). Intrusion detection based on distance combination. International Journal of Computer Science, 2(3), 178–186.
4. Kim, G., Lee, S., Kim, S. (2014). A novel hybrid intrusion detection method integrating anomaly detection with misuse detection. Expert Systems with Applications, 41(4), 1690–1700.
5. Depren, O., Topallar, M., Anarim, E., Ciliz, M. K. (2005). An intelligent intrusion detection system for anomaly and misuse detection in computer networks. Expert Systems with Applications, 29(4), 713–722.
6. Luo, B., Xia, J. (2014). A novel intrusion detection system based on feature generation with visualization strategy. Expert System with Applications, 41, 4139–4147.
7. Lin, S. W., Lee, Z. J., Chen, S. C., Tseng, T. Y. (2008). Parameter determination of support vector machines and feature selection using simulated annealing approach. Applied Soft Computing, 8(4), 1505–1512.
8. Mukherjee, S., Sharma, N. (2012). Intrusion detection using Naïve Bayes classifier with feature reduction. Procedia Technology, 4, 119–128.
9. Lin, S. W., Ying, K. C., Lee, C. Y., Lee, Z. J. (2012). An intelligent algorithm with feature selection and decision rules applied to anomaly intrusion detection. Applied Soft Computing, 12(10), 3285–3290.
10. Wu, X., Kumar, V., Quinlan, J. R., Ghosh, J., Yang, A., Motoda, Y., McLachlan, G. J., Ng, A., Liu, B., Yu, P.S. (2008). Top 10 algorithms in data mining. Knowledge and Information System, 14(1), 1–37.
11. Yang, J., Olafsson, S. (2006). Optimization-based feature selection with adaptive instance sampling. Computer & Operation Research, 33(11), 3088–3106.
12. Tavallaee, M., Bagheri, E., Lu, W., Ghorbani, A.A. (2009). A detailed analysis of the KDD Cup data sets. In Prococeedings of the 2nd IEEE Symposium on computational intelligence in security and defense applications (pp. 53–58).
13. KDD Cup'99 Data, http://kdd.ics.uci.edu/databases/kddcup99/kddcup99.html
14. Quinlan, J. R. (1986). Introduction of decision trees. Machine Learning, 1, 81–106.
15. Quinlan, J. R. (1987). Decision trees as probabilistic classifiers. In Proceedings of the 4th International Workshop Machine Learning (pp. 31–37).
16. Quinlan, J. R. (1993). C 4.5: programs for machine learning. San Mateo: Morgan Kaufmann Publishers.
17. Quinlan, J. R. (1996). Learning decision tree classifier. ACM Computing Surveys (CSUR), 28 (1), 71–72.
18. Chang, C., Lin, C. (2011). LIBSVM: A library for support vector machines. ACM Transactions on Intelligent Systems and Technology, 2(3), 27:1–27:27. Software available at http://www.csie.ntu.edu.tw/~cjlin/libsvm
19. Vapnik, V. (1995). The Nature of Statistical Learning Theory. Springer-Verlag, New York.
20. Schölkopf, B., Platt, J. C., Taylor, J. S., Smola, A. J., Williamson, R. C. (2001). Estimating the support of a high-dimensional distribution. Neural Computation, 13(7), 1443–1471.
21. Perdisci, R., Gu, G., Lee, W. (2006). Using an ensemble of one-class SVM classifiers to harden payload-based anomaly detection systems. In Proceedings of the 6th International Conference on data mining (pp. 488–498).

22. Hall, M., Frank, E., Holmes, G., Pfahringer, B., Reutemann, P., Witten, I. H. (2009). The WEKA data mining software: An update. ACM SIGKDD Explorations Newsletter, 11 (1), 10–18.
23. Song, J., Takakura, H., Okabe, Y., Kwon, Y. (2009). Unsupervised anomaly detection based on clustering and multiple one-class SVM. IEICE Transactions on Communications, E92-B (6), 1982–1990.

Analysis of Reconfigurable Fabric Architecture with Cryptographic Application Using Hashing Techniques

Manisha Khorgade and Pravin Dakhole

Abstract Coarse-Grained Reconfigurable Architecture (CGRA) is currently receiving attention as it is a strong emerging class with excellent performance as well as flexibility in fabrication. System building blocks uses the entire range of components is available as choices for. The Reconfigurable fabric (RF) will be offered as an important building block for complex system design, CGRA processor. This paper gives an innovative design of reconfigurable fabric (RF) performing with parallel processing techniques. Cryptographic hash function is a hash function which cannot be inverted practically, to regenerate the input data from its hash value alone. RF having 16 processing elements (PE) in mesh-type topology for single dimensional processing of encryption technique. The analyzing parameters for this design are power, processing speed and area on various FPGAs.

Keywords Reconfigurable fabric · Crypto graphical application · Hashing techniques · Interconnectivity mesh · FPGA

1 Introduction

As recent scenario the complexity has increased a far, which is driving designers in wireless and multimedia applications to innovate constantly. Energy–area–timing are fundamental trade-off between flexibility and the traditional performance metrics always. The building blocks of modern System-on-Chip (SoC) in today's era are fetching attention with the balance of performance and flexibility. Coarse-Grained Reconfigurable Architecture (CGRA) is the one key building block which is having strong performance advantage and also ability of flexed post fabrication. Recent design proposals for both academic and commercial include CGRAs for DSP applications [1–3] or are completely based upon CGRAs [4, 5]. A Coarse-Grained Reconfigurable Architecture (CGRA) is a processing platform

M. Khorgade (✉) · P. Dakhole
YCCE, Nagpur, India
e-mail: manisha.khorgade@gmail.com

© Springer Nature Singapore Pte Ltd. 2017
S. Patnaik and F. Popentiu-Vladicescu (eds.), *Recent Developments in Intelligent Computing, Communication and Devices*, Advances in Intelligent Systems and Computing 555, DOI 10.1007/978-981-10-3779-5_11

which constitutes an interconnection of coarse-grained computation units, processing elements (PEs) and arithmetic logic units (ALUs). These units communicate directly are used in multi-core processors. CGRAs are a well-researched topic, and the design space of a CGRA is quite large. The design space can be represented as choice of computation unit, choice of interconnection network, Choice of number of context frame (single or multiple), presence of partial reconfiguration, choice of orchestration mechanism, design of memory hierarchy and host-CGRA coupling. Restrictions to mostly traditional application of FPGAs are prototyping and emulation, and design methodology of CGRAs which is again the emerging field. In spite of approach followed by these design flows, the fundamental challenges of design persist. A fully novel approach for exploring the efficient units of a RF using mesh-based topology in CGRA with corresponding mapping algorithm is presented in this paper. The main part of RF unit is functional unit. It is like ALU. In this designing approach, it is considered as processing element (PE). There are different types of placement and routing techniques used for reconfigurable fabric unit and processing element. In this paper, we will discuss execution of RF using mesh-type topology to analyze area, time, throughput and speed for edge detection technique. This paper contains methodology for implementation in Sect. 2; similarly, Sects. 3 and 4 will give implementation of PE and RF unit. Section 5 contains operation of system and analysis which is studied, and results are contained in Sect. 6. Section 7 contains conclusion and after that references.

2 Methodology

The software and hardware techniques are used to implement reconfiguration system of DSP processor architecture. Basically, it has two types such as coarse grained (CGRA) and fine grained (FGRA). CGRA operates with bytewise input and FGRA operates with bitwise input. They are having their own advantages and disadvantages. Our approach towards CGRA with reconfigurable fabric (RF) also control unit, IO unit.

The RF comprises with set of processing elements (PE). Topology used to connect all these PEs is mesh type. All PEs designed like homogeneous granularity. Broadly, they accept n-bit stream inputs with n-bit stream outputs. There are other types of topologies exist for PEs arrangement as bus and crossbar. The most important task is to schedule given task and distribute it between PEs to execute. During scheduling, record is maintained to keep track of which resources are used in every time period. In view of scheduler for a CGRA, it must perform operations of placement, routing information. The information is included to record, since PEs are used for computation along with routing. Organization of interconnect network is not necessary, since connections are dedicated as point-to-point connections. This indicates that there is no blocking can occur in the network. In order to that, parallelism is achieved with scalability through a range of applications (Fig. 1).

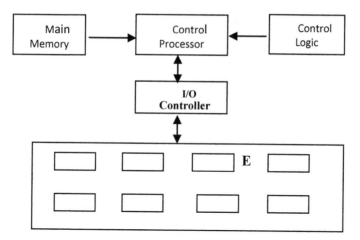

Reconfigurable Processing Unit(RPU)

Fig. 1 Reconfigurable processor architecture

Here, task has been distributed and feed into 2 * 8 arrangement of PEs to accomplish parallelism. For observing outcome of such arrangement, different applications like hashing method using xoring technique are presented.

3 Implementation of Reconfigurable Fabric

Reconfiguration computing includes multiple PEs arrangement with variety of topologies performing diverse tasks at same time. Selection of reconfiguration model is done prior to run program, so that reconfiguration technique is considered to be static reconfiguration. For static reconfiguration computing, we have implemented 2 Nodes as controlling units, 4 PEU as reconfigurable processing unit with 8 PEs, and to control logic are designed. The granularity is coarse grain type including topology—mesh type shown in Fig. 3a, b also. As in CGRA designing, in mesh-based architectures, PEs are arranged in an array, featuring horizontal and vertical connections. As a result, this formation allows efficient parallelism, furthermore excellent utilization of communication resources. Still the advantages of a mesh are intense for the need of an efficient placement and routing step. The quality of this step can have a remarkable impact on the application performance. PEs used granularity as homogeneous type i.e. even granularity (Fig. 2).

Depending on the requirement of PEs as per task, scale-in and scale-out of unit decision taken by control node. With the help of this control unit, various applications can be run simultaneously, so with this reconfiguration is also achieved.

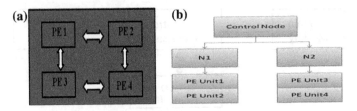

Fig. 2 **a** Mesh-type arrangement of RF using PEs, **b** design of proposed RF Unit using 2 Core * 8 PEs Arragement

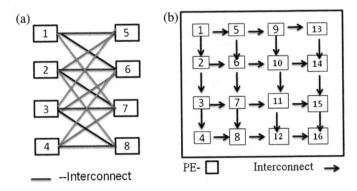

Fig. 3 Interconnectivity topologies as **a** crossbar, **b** mesh type

4   Interconnect Topologies

Since there are various types of interconnect topologies in reconfigurable computing. Communication between processors, their caches along memory is the standard i.e. shared-bus interconnect. Disadvantage of this technique is that there are significant conflict on the interconnect, because one processor can use interconnect at significant given time. This is the motivation focussed on additional interconnect topologies for the possibility of optimizing different parameters of its design—for example, it may select to optimize the latency or scalability of an interconnect system, or the actual cost of its real-world implementation might be significant for optimizing as a substitute.

A. Mesh type

As shown in Fig. 4b, each node is connected to its own switch as 2D grid of nodes, this is called as the 2D mesh network topology. Here, frequency of sending a message from one node to another node is non-uniform. The latency on average is O (sqrt(N)). As there is dedicatedly one switch per node is allotted, the cost is O(N). This topology is improved than the crossbar network topology but the clear trade-off is advanced latency.

Fig. 4 Delay calculation for mesh-type topology

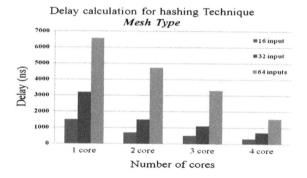

Delay calculation for hashing Technique
Mesh Type

Table 1 Bitwise distribution with controller, PEU and PE selection

Bit	Bit value	Controller	PEU	PEs
B2	0	C1	–	–
	1	C1 and C2		–
B1	0	C2	PEU0	2
	1	C2	PEU0, PEU1	4
B0	0	C1	PEU0	2
	1	C1	PEU0, PEU1	4

Second characteristics, the layout is very easy to design. So chip designing is also effective. It is measured direct as nodes are connected to switches which are inside the network itself. Several ways are there for messages to pass through from one node to the other. Still, the mesh topology is considered as best topology.

5 Operation

It has 3-bit operation mode $(B_2B_1B_0)$ to perform depending upon the values of bits from MSB (B2)–LSB (B0). For selection of controller C1 and C2, B2 (MSB) is selected. It means value of MSB decides which controller is selected. If the value of B2 is 0, controller C1 is selected and if B2 is 1, Controller C1 and C2 are selected. C1 has 2 PEU. Each PEU has 2 PES, similar to C2 also. So if C1 selected, it will select 2 PEU with 4 PEs. If C1 and C2 are selected, it will select 4 PEU with 8 PEs. If B1 bit is selected and bit is 0, C2 is selected which select PEU0 (2 PEs). If B1 is 1, C2 is selected which select PEU0 and PEU 1 (4 PEs). For third bit (B0), it will select controller C1. If B0 is 0, C1 is selected with PEU0 (2 PEs) and if B0 is 1, it will select PEU0 and PEU1 (4 PEs) (Table 1).

The table shown below gives the core distribution for the input bit as explained in operation. As there are four cores, the allotted task is scheduled and distributed to the respective core and that will be get executed. For selection of one core delay is more and if four cores are selected, delay automatically get reduced (Table 2).

Table 2 Relationship of number of cores and task performed

Sr No	Input bit	No of cores selected	Task performed	
1	00	1	Complete task	100%
2	01	2	Task divided in 2 equal parts	50%—core 1 50%—core 2
3	10	3	Task divided in 3 equal parts	1/3 part—core 1 core 2, core 3
4	11	4	Task divided in 4 equal parts	¼ part—core 1, core 2 core 3 core 4

6 Results

The approach is to design and analyze cryptography application with hashing technique on various parameters. Analysis is done for a range of bit streams sizes such as 16, 32, 64 bits. Several parameters are considered for evaluations such as time, area, power, speed and throughput. Power and speed (signal).

6.1 Hashing Function

A cryptographic hash function is a hash function. This is practically not possible to reverse, i.e. to restore the input data from its hash value only. The input data is called the message, and the hash value is often called the message digest or simply the digest. The ideal cryptographic hash function has four main properties; as it is easy to compute the hash value for any given message, it is impossible to generate a message from its hash also not possible to alter a message without varying the hash. It is used to index the original value or key and then uses later each time the data associated with the value or key is to be retrieved. So, hashing is always a one-way operation. It cannot be done using "reverse engineer" the hash function by analyzing the hashed values. Indeed, the ideal hash function cannot be derived by such analysis. A hash function is called as good hash function which should not produce the same hash value with two different inputs. If this happens, it is known as a collision.

The input to the proposed architecture changes from 16-bit input, 32-bit input and 64-bit input. As we are changing number of cores from one core to four cores, respectively, the delay time and accordingly speed also changes for the type of mesh-type interconnecting topologies as shown in Table 3.

Also, we have tried to calculate area occupied by the proposed architecture of reconfigurable fabric using FPGA. While dumping on the FPGA, we have selected Spartan 3E and Spartan 6 kit are selected [6–8]. This family give the area covered with number of slice register, flip flops, LUTs, IOBs and maximum number of path covered i.e. wire length (Table 4).

Table 3 Number of cores used and delay taken for processing

Core used	Cryptography mesh-type delay (ns)		
	Input 16 bits (ns)	32 bit (ns)	64 bit (ns)
00	1505	3200	6567
01	700	1492	4739
10	505	1097	3318
11	305	700	1548

Table 4 Area calculations: cryptography with mesh type arrangement

S. no	Parameter	Cryptography Application for (16 Inputs), (32 inputs)			
		Spartan 6	Spartan 3E	Spartan 6	Spartan 3E
1	Slice register	337	333	469	2735
2	Flip-flop	1027	333	1796	461
3	LUTs	1069	2718	1779	5230
4	IOBs	261	261	517	517
5	Min period (ns)	5.918	18.179	5.948	18.615
6	Max frequency (MHz)	168.98	55.009	168.136	53.72

Here, we have application for hashing techniques such as verifying the integrity of files or messages, password verification, proof-of-work and also file or data identifier.

7 Conclusion

The novel approach of reconfigurable fabric design is presented. Especially, it has been noticed that reconfigurable computing is grown at the same time as a large discipline. Hence, it demands separate attention to the reconfigurable processors designing area. There are various techniques for designing and optimizing different parameters have been proposed earlier. Still, the effective design of reconfigurable computing architecture is a big challenge. Our approach is to give novel set of arrangement of reconfigurable fabric to optimize the trade-off between area, power and efficiency. We have implemented this architecture and analyzed it using one of the application, as hashing techniques in cryptography. Here, analysis also done for speed and throughput of proposed model. Anticipating the future challenges, several research directions are optimization of PE, RF with these parameters and many others. It is likely that our understanding of SoC architectures will evolve with time. In near future with proposed model, we will implement the application for image/video coding–decoding to achieve scalability with reconfiguration.

References

1. "High-level Modelling and Exploration of Coarse-grained Re-configurable Architectures", Amupam Chattopadhyay, Design, Automation and Test in Europe, DATE 2008, Munich, Germany, March 10–14, 2008
2. Hyunchul Park, Kevin Fan, "Modulo Graph Embedding: Mapping Applications onto Coarse Grained Reconfigurable Architectures", Proceedings of the 2006 - dl.acm.org/
3. "Low Power Reconfiguration Technique for Coarse-Grained Reconfigurable Architecture", Yoonjin Kim, Rabi N. Mahapatra, Ilhyun Park, IEEE TRANSACTIONS ON VERY LARGE SCALE INTEGRATION (VLSI) SYSTEMS, VOL. 17, NO. 5, MAY 2009
4. "Design Space Exploration for Efficient Resource Utilization in Coarse-Grained Reconfigurable Architecture", Yoonjin Kim, , Rabi N. Mahapatra, Ilhyun Park, IEEE, IEEE TRANSACTIONS ON VERY LARGE SCALE INTEGRATION (VLSI) SYSTEMS, VOL. 18, NO. 10, OCTOBER 2010
5. "A Design Flow for Architecture Exploration and Implementation of Partially Reconfigurable Processors" Kingshuk Karuri, Anupam Chattopadhyay, Xiaolin Chen, David Kammler, Ling Hao, Rainer Leupers, IEEE TRANSACTIONS ON VERY LARGE SCALE INTEGRATION (VLSI) SYSTEMS, VOL. 16, NO. 10, OCTOBER 2008
6. "Reducing Control Power in CGRAs with Token Flow", Hyunchul Park, Yongjun Park, and Scott Mahlke, DATE 2008
7. Allan Carroll, Stephen Friedman, "Designing a Coarse-grained Reconfigurable Architecture for Power Efficiency", proceeding 8 Proceedings of the ACM/SIGDA international symposium on Field programmable gate arrays, pages 161–170
8. "A Coarse Grained Reconfigurable Architecture Framework supporting Macro-Dataflow Execution", thesis submitted Keshavan Varadarajan, december 2012 at IIS, Bangalore

Privacy Preservation of Infrequent Itemsets Mining Using GA Approach

Sunidhi Shrivastava and Punit Kumar Johari

Abstract Privacy preservation of information is an important approach to data mining. Infrequent or rare itemset mining is a new technique in this field which is very useful for gaining profit from the business point of view. Rare thing can make more profit. Misuse of these techniques can lead to revelation of confidential information. In this paper, we addressed this problem of privacy preservation of data mining by using sanitization of database or in the other word hiding high utility rare itemsets. We have identified high utility rare patterns and introduce an approach for dynamic addition of transactions. The central goal of the proposed algorithm is to optimize high utility rare items for providing privacy.

Keywords Privacy preservation · Data mining · Rare itemset mining · Utility mining · Genetic algorithm

1 Introduction

Data mining is a methodology of mining valuable or meaningful information using various mining techniques [1]. The objective of data mining is to retrieve higher-level indiscernible information from a plenty of raw data. Enlarging network complexity, providing greater access, sharing information, and a prospering emphasis on the Internet have caused information security and privacy a notable concern for human beings and organizations [2]. In data mining, association rule mining (ASM) techniques identify the profit value of any itemset by its occurrence in the transaction set [3]. All the itemsets having utility higher than some threshold

S. Shrivastava (✉) · P.K. Johari
Madhav Institute of Technology and Science, Gwalior, India
e-mail: sunidhishrivastava5@gmail.com

P.K. Johari
e-mail: punitbhopal2006@gmail.com

© Springer Nature Singapore Pte Ltd. 2017
S. Patnaik and F. Popentiu-Vladicescu (eds.), *Recent Developments in Intelligent Computing, Communication and Devices*, Advances in Intelligent Systems and Computing 555, DOI 10.1007/978-981-10-3779-5_12

will be considered as high utility items, and this whole process is called as utility mining [4, 5]. Like, smartphone can be more profitable than the normal phone [6, 7]. Itemset that occurs frequently is called frequent itemsets, and itemset that does not occur frequently in the database is called infrequent items in the database ([8, 9]). The genetic algorithm is used to provide better optimization standard in the field of evaluation and privacy [10, 11].

2 Literature Survey

H. Yao et al. proposed [12] the utility problem-based mining to discover the itemsets that are noteworthy according to their utility values. J. Hu et al. [13] proposed an algorithm based on the frequent itemset mining technique for discovering high utility pattern. In disparity of the classical association rule and frequent item mining methods, the aim of the algorithm is to find data segments defined by the few items' (rules) groupings, which fulfill various situations present in an effective estimate to solve it by particular partition trees, known as high-yield partition trees. Jyothi Pillai et al. [14] proposed Apriori inverse is used to find only the rare itemsets. HURI is used to find those rare itemsets, which are of high Utility according to users' preferences, i.e., algorithm for generation of rare itemsets is extended to find high-utility rare itemsets. It means user will setup some criteria for high utility items and set Max high utility value according that High Utility Rare Itemsets will be identify. Hence, HURI is said to be more helpful on the application of a real-world dataset. In future, the algorithm can be worked with the fuzzy and temporal concept for mining high profitable itemset. Tseng et al. [15] proposed a new technique specifically temporal high utility itemsets (THUI) for timing-based HUI from databases effectively and efficiently. S.A.R. Niha et al. [16] proposed UPRI algorithm for generating high utility rare itemsets. UPR tree has been used for representing information in the tree-based structure. They have applied four strategies in the process of extracting high utility patterns of the itemsets. High utility patterns are those all set of items which have high utility while together.

3 Problem Definition

Privacy preservation of data mining always focuses on the frequent itemsets, where highly profitable rare itemsets are left out. So we focused on infrequent itemsets.

4 Proposed Framework

The proposed framework is shown in Fig. 1.

Fig. 1 Proposed flowchart showing the privacy preservation of high utility rare itemset

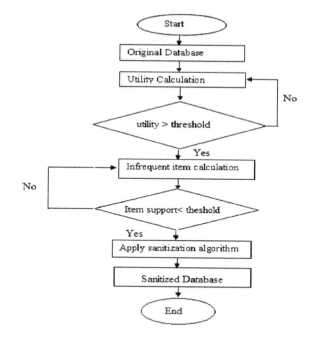

5   Proposed Algorithm

Input:

1. A transactional dataset,
2. A profit dataset,
3. Min_support threshold,
4. Min_utility threshold, and
5. Dynamical addition of transactions.

Output: A sanitized database with no sensitive itemsets.

Method: In the proposed methodology on the basis of two user specified threshold, we have find out high utility rare itemsets (HURI).

1. **Min_utility threshold**: In utility mining, minimum utility is a user-defined value, and itemsets having a value greater than the min_utility threshold is considered as a high utility itemset.
2. **Min_support threshold**: A frequent itemset is the itemset containing frequency support higher than a minimum user-specified threshold and the rare itemset which has the support lower than the minimum user specified threshold.

The propose method is comprised of the following steps:

1. First, calculate the transaction utility of each transaction, transaction-weighted utility, and item's utility of each item in transactions.
2. Apply minimum utility threshold on transaction-weighted utility, and items having lesser value will be discarded.
3. Apply minimum support threshold, and items having lesser value will be extracted as rare itemset. So, finally, high utility rare itemsets are received from the transaction.
4. Apply genetic algorithm for hiding high utility rare items.
5. Take maximum length itemset from the high utility rare itemsets.
6. Use the point crossover for flipping the values from the database.
7. Sanitize the database.
8. At the end, calculate the total elapsed time.

5.1 Illustrated Example

In this section, an example is given to show the proposed algorithm. First, we will take a database shown in Tables 1 and 2. A user-specified threshold is set to min_utility = 50 and min_supp = 0.7.

Table 1 consists of seven items and five transactions. And Table 2 shows the profit value associated with all transactions. To identify high utility rare itemsets, the following formulas will be used.

Item's utility: An item's utility will be implemented by taking the product of profit and items value.

$$U(A) = T1(A) + T2(A) + T3(A) + T5(A)\ldots \ast \text{profit}(A) \tag{1}$$

Table 1 Database with transactions

Transactions	A	B	C	D	E	F	G
T1	1	1	2	1	0	0	0
T2	2	0	5	0	2	4	5
T3	3	1	1	5	1	4	0
T4	0	5	2	2	1	0	0
T5	1	1	2	0	1	0	2

Table 2 Utility associated with items

Product	A	B	C	D	E	F	G
Utility	5	7	3	1	3	4	2

Table 3 New database

Transactions	A	B	C	D	E	F	G
T1	1	0	0	1	0	0	0
T2	1	0	1	0	2	1	1
T3	0	0	1	5	1	0	0
T4	0	1	0	2	1	0	0
T5	1	1	2	0	1	0	2

Utility: Now, we can also calculate the transaction utility by multiplying each item with their utility from 1 transaction at one time.

$$TU(T1) = A(T1) * P(A) + B(T1) * P(B) + C(T1) * P(C) + D(T1) * P(D)$$
$$+ E(T1) * P(E) + F(T1) * P(F) + G(T1) * P(G) \tag{2}$$

Transactions-weighted utility: Summation of all items' transaction utility is called as transaction-weighted utility.

$$TWU(A) = TU(T1) + TU(T2) + TU(T3) + TU(T5) \tag{3}$$

By using formulas 1, 2, and 3, we will get the total utility value of all the items, and after calculating these utilities, we will compare all the *TWU* with min_utility threshold, itemsets having value larger than the support considered as higher utility itemsets. After calculating the support of all HUI, in the next step, we will generate all the itemsets which have the support value less than the min_supp threshold.

By using this approach, we have high utility rare itemsets such as *D, F, G, DF, DG, FG, DFG*.

Now, from the process of genetic algorithm, first, maximum length itemset will be considered, that is, *DFG*. Crossover operation is performed using the genetic algorithm. Iterations are the total number of loop or attempts which are required for extracting all the high utility rare itemsets are identified. Here 11 iterations are required (Table 3).

After this whole process, no high utility rare itemset will be visible from the database, and now, we have a highly confidential database.

6 Result Analysis

For the analysis, we have used graph that shows the difference between original and sanitized items. The graph is plotted by taking occurrence of items at the x-axis and the number of transactions on the y-axis. From the graph, it can easily identify that

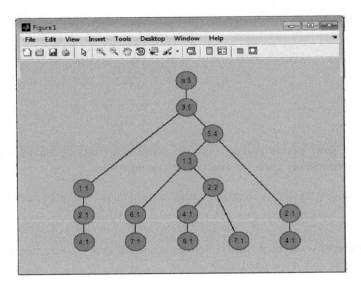

Fig. 2 UPR tree

Fig. 3 Item *A* before and after sanitization process

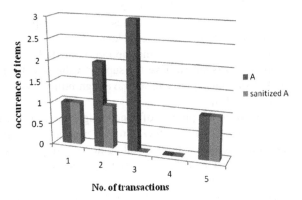

all the high utility rare itemsets are hidden. So by using the proposed algorithm based on two threshold values, all the HURI will be optimized.

1. To understand the process of generating infrequent itemset, the following graph which is called UPR tree is used (Fig. 2).
2. Figures 3 and 4 show the difference between original item and sanitized item, after applying privacy preservation. The difference between all the items present in the database can be shown like this. Total execution time of the whole process is 2.327772 s.

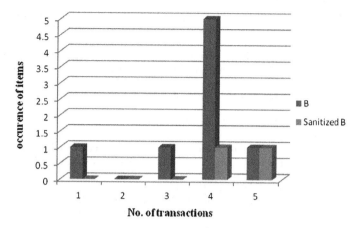

Fig. 4 Item *B* before and after sanitization process

7 Conclusion

Extracting meaningful information plays an important role in the mining process. More accurate data can give better result. Privacy preservation of profitable items is also necessary. In this paper, a novel framework has been implemented for privacy preservation of infrequent itemset mining. High utility rare itemsets are identified, and dynamic approach to adding transactions into the database also introduced. We have introduced an efficient implementation of genetic algorithm for providing privacy preservation of infrequent itemsets.

References

1. Jain, Nikita, and Vishal Srivastava. "Data Mining techniques: A survey paper." IJRET: International Journal of Research in Engineering and Technology 2.11 (2013): 2319–1163.
2. Janakiramaiah, B., A. Rama Mohan Reddy, and G. Kalyani. "Privacy Preserving Frequent Itemset Mining by Reducing Sensitive Items Frequency using GA".
3. Joshi, Maya, and Mansi Patel. "A Survey on High Utility Itemset Mining Using Transaction Databases." Vol. 5 (6), 2014, 7407–7410.
4. Tseng, Vincent S., et al. "Efficient algorithms for mining high utility itemsets from transactional databases." Knowledge and Data Engineering, IEEE Transactions on 25.8 (2013): 1772–1786.
5. Lin, Jerry Chun-Wei, et al. "Efficient algorithms for mining high-utility itemsets in uncertain databases." Knowledge-Based Systems (2016).
6. Sudip Bhattacharya, Deepty Dubey, "High Utility Itemset Mining", International Journal of Emerging Technology and Advanced Engineering, ISSN 2250-2459, Volume 2, Issue 8, August 2012.
7. H. Yao and H. J. Hamilton, "Mining itemset utilities from transactio databases," Data and Knowledge Engineering, vol. 59, pp. 603–626 2006.

8. Endu Duneja and A.K. Sachan, "A Survey on Frequent Itemset Mining with Association Rules", International Journal of Computer Applications (0975 – 8887) Volume 46– No. 23, May 2012.
9. Sethi, Nidhi, and Pradeep Sharma. "Efficient Algorithms for Mining Rare Itemset over Time Variant Transactional Database." International Journal Of Computer Science and Information Technologies 5 (2014).
10. Srinivas, Mandavilli, and Lalit M. Patnaik. "Genetic algorithms: A survey." Computer 27.6 (1994): 17–26.
11. R.K. Battacharya, Introduction to Genetic Algorithm, Indian Institute of Technology, Guwahati, 2012.
12. Hong Yao, Howard J. Hamilton, Liqiang Geng, "A Unified Framework for Utility Based Measures for Mining itemsets", In Proc. of the ACM Intel. Conf. on Utility-Based Data Mining Workshop (UBDM), pp. 28–37, 2006.
13. Jianying Hu, Aleksandra Mojsilovic, "High-utility pattern mining: A method for discovery of high-utility item sets", Pattern Recognition 40 (2007) 3317–3324.
14. Jyothi Pillai, O.P. Vyas, "High Utility Rare Itemset Mining (HURI): An Approach for Extracting High-Utility Rare Item Sets." Journal on Future Engineering and Technology 7, no. 1 (2011).
15. Vincent S. Tseng, Chun-Jung Chu, Tyne Liang, "Efficient Mining of Temporal High Utility Itemsets from Data streams", Proceedings of Second International Workshop on Utility-Based Data Mining, August 20, 2006.
16. S.A.R. Niha, Dr Uma N Dulhare, "Extraction of High Utility Rare Itemsets from Transactional Databases." Computer and Communications Technologies (ICCCT), 2014 International Conference on. IEEE, 2014.

A Quinphone-Based Context-Dependent Acoustic Modeling for LVCSR

Priyanka Sahu and Mohit Dua

Abstract Automatic speech recognition (ASR) is used for accurate and efficient conversion of speech signal into a text message. Generally, speech signal is taken as input and it is processed at front end to extract features and then computed at back end using the GMM model. GMM mixture selection is quite important depending upon the size of dataset. As for concise vocabulary, use of triphone-based acoustic modeling exhibits good results but for large size vocabulary, quinphone (quadra-phones)-based acoustic modeling gives better performance. This paper compares the performance of context-independent- and context-dependent-based acoustic modeling to reduce error rate.

Keywords Speech recognition · Speech modeling · Feature extraction techniques · Gaussian mixture model · LVCSR

1 Introduction

The reformation of speech signal wave into its corresponding text (a helpful message) is called automatic speech recognition (ASR). ASR demands the well-organized compression of inserted speech material to a tiny set of parameters, having the power to perfectly categorize sections of data as phonemes. In the absence of data reduction, categorization of speech to text is not realistic due to the large number of possible inserted audio waves and the inadequate way to link up all such type of waves to their respective text. System should be independent of environmental noise, speaker speaking style, accent, along with the property to handle the large size of vocabulary [1]. State-of-art ASR system follows the statistical pattern recognition approach that includes feature extraction at the front end

P. Sahu (✉) · M. Dua
National Institute of Technology Kurukshetra, Kurukshetra, Haryana, India
e-mail: er.priyankasahu40@gmail.com

M. Dua
e-mail: er.mohitdua@gmail.com

© Springer Nature Singapore Pte Ltd. 2017
S. Patnaik and F. Popentiu-Vladicescu (eds.), *Recent Developments in Intelligent Computing, Communication and Devices*, Advances in Intelligent Systems and Computing 555, DOI 10.1007/978-981-10-3779-5_13

and likelihood evaluation of features at back end. However, ASR has been made significant advancement in legible applications such as dictation and medium vocabulary transaction processing acts. Gaussian mixture computation of acoustic signals is an expensive task of computation. In such system, calculation of state likelihoods makes a significant proportion (between 30 and 70%) all over the computational load [2]. A range of 8–64 mixture components per state have been found useful depending on the amount of training data.

2 Working of ASR

The motive of automatic speech recognition is to make a machine intelligent to "listen," "perceive," and then "respond" to the speech. Automatic speech recognition systems can be broadly classified into two modules: (1) front-end module (includes preprocessing and feature extraction), and (2) back-end module (includes modeling and pattern classification). Figure 1 shows a statistical framework for speech recognition.

Preprocessing is the task to convert analog signal into numerical values for digital processing. Preprocessing includes the following: background noise removal, pre-emphasis, framing, and windowing [3, 4]. Feature extraction transforms the waveform of speech into some suitable parametric relations that are further used to analyze the speech-related applications such as speech coding and speaker identification. MFCC and LPCC are the two commonly used feature extraction techniques [5]. Knowledge model is basically composed of three submodels; these are acoustic model, language model, and lexicon model. Its major role is to map the speech features into proper word order. Acoustic modeling is used to identify speech phoneme.

Hidden Markov model (HMM) is the most efficient and common tool used for acoustic modeling; probability evaluation, best sequence determination, and

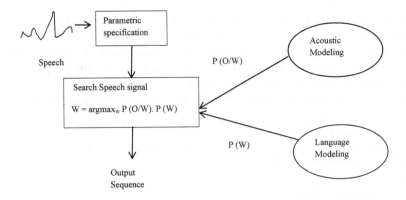

Fig. 1 Statistical speech recognition framework

estimation of parameters are the three basic problems of HMM. Forward algorithm is used to realize probability density function; Viterbi search and A* search are used for best sequence determination; and parameter estimation is realized using maximum likelihood estimation (MLE) [6]. Language model is used to process the output of acoustic model. Intuitively, it calculates the probability of an utterance taking place after the set of words. N-gram language modeling (lying on prophecy of nth string on the basis of previous $n - 1$ word strings) is the most common practice in automatic speech recognition. Lexicon modeling gives clarification of words in sense of fundamental phonemes in system vocabulary means it give pronunciation of each vocabulary word. Pattern classification is the act of recognition of sample words based on its acoustic properties. Most probable word sequence got produced after the feature vector decoding.

3 Phonetic Representation of Speech Signals

In automatic speech recognition system, the representation of speech and non-speech units is vital. Phones get co-articulated when a phone overlaps with the margins of its neighboring phone. So in order to avoid this co-articulation effect, context-dependent modeling is required to develop. To avoid the effect of co-articulation, the acoustic realization of a particular phoneme is adversely dependent on its previous and upcoming next phonemes and this is known as phonetic context. There are three context-dependent phones in use: monophones, triphones, and quinphones. Context dependency (CD) is essential if high recognition accuracy is targeted.

3.1 Modeling Using Monophones

The monophones are single phones; the phonetic transcription of the word FATHER is given as follows:

$$BHARATIYA \quad /bharatiya/ \tag{1}$$

As vocabulary size is small, monophones are good in result, but as dataset size increases, triphones and quinphones are used. For large vocabulary continuous speech recognition, whole word model implementation is totally infeasible because in that case, several of dozens of realization are made for each word. So for large vocabulary datasets, triphones and quinphones are developed in order to improve the accuracy of Hindi speech recognition.

3.2 Modeling Using Triphones

Triphones are quite common, where each subwords are phonemes in the context of left and right phonemes. These phones have a window size of three phonemes. It can be represented by $A - B + C$, where phone B is in left context to A and in right context to C. Context-dependent models such as triphones can be constructed either by internal word or by crossword. In internal word models, the context beyond the boundaries of word is not well thought out. On the other side, in crossword phonemes, phonemes present at the beginning or end of neighboring words are involved to affect the phonology used for modeling. For example, BHARATIYA is a word and its triphone phonetic transcription is as follows:

$$bh(*, a)a(bh, r)r(a, a)a(r, t)t(a, i)i(t, y)y(i, a)a(y, *) \tag{2}$$

3.3 Modeling Using Quinphones

Quinphones have wider level of phonetic context than monophones or triphones. For very large vocabulary size, triphone modeling degrades the accuracy, so quinphones (quadraphones) are used. In the case of quinphones, there are two neighboring phones (phonemes) present in the context of preceding and following phonemes. These phones contain window length of five phones. The phonetic representation for quinphones of the word BHARATIYA is as follows:

$$bh(*, ar)a(bh, ra)r(bha, at)a(ar, ti)t(ra, iy)i(at, ya)y(ti, a)a(iy, *) \tag{3}$$

Context-dependent modeling using quinphones reduces this class of model and allows the large rise in the computation to be tolerated. Modeling using quinphones and triphones would be difficult if the basic number of phones is too big. For example, let there be 40 phones, then its triphones are already of the order of 40^3 and quinphone combinations may be greater than 64000 phones [7, 8]. One way to reduce this dimensionality is to cluster the triphones based on acoustic features.

4 Implementation

On the basis of speaking mode, automatic speech recognition is categorized into four parts: (a) isolated word recognition, (b) connected word recognition, (c) continuous speech recognition, and (d) spontaneous speech recognition. In this paper, automatic speech recognition system is developed for continuous speech recognition for Hindi language. HTK 3.4.1 toolkit from Cambridge University is used for implementing the ASR on Linux (Ubuntu 14.04) platform.

4.1 Speech Data Preparation

In order to develop any ASR system, its first obvious step is data preparation. Speech data is needed for training and testing the system. In this paper, 10–1000 words of speech data are considered for training. These speech datasets are taken from IIT Hyderabad open source speech datasets.

4.2 Training

During training phase, HMM model is fabricated. This HMM prototype is generated for each and every Hindi monophone including silence model. Monophones, triphones, and quinphones are used to train the system as vocabulary size increases from few to thousands of words, respectively. Figure 2 displays few quinphones produced during experiments.

4.3 ASR Performance Analysis

The accuracy is evaluated by word error rate (WER), which means recognizing the total incorrect words. Word error rate is also observed in smaller units such as phonemes and syllables. WER can be calculated as follows:

$$\text{WER} = (S + D + I)/N \tag{4}$$

where S (substitution rate), D (deletion rate), I (insertion rate) are some more detailed errors and N is number of words in the reference. Second criterion speed is measured using real-time factor and requires time P to process an input of I duration,

$$\text{RTF} = P/I \tag{5}$$

```
CL quinphones1
TI  T_aa {(*^aa-aa+aa=*,*^aa-aa+*,*-aa+aa=*,*^aa-*,*-aa+*,aa+*,+aa=*,*-
aa).transP}
TI T_j {(*^j-j+j=*,*^j-j+*,*-j+j=*,*^j-*,*-j+*,j+*,+j=*,*-j).transP}
TI T_a {(*^a-a+a=*,*^a-a+*,*-a+a=*,*^a-*,*-a+*,a+*,+a=*,*-a).transP}
TI T_k {(*^k-k+k=*,*^k-k+*,*-k+k=*,*^k-*,*-k+*,k+*,+k=*,*-k).transP}
TI T_l {(*^l-l+l=*,*^l-l+*,*-l+l=*,*^l-*,*-l+*,l+*,+l=*,*-l).transP}
```

Fig. 2 Some quinphone entries

Table 1 Performance evaluation of ASR for different vocabulary sizes

Vocabulary sizes	Monophones (context-independent modeling) (%)	Triphones (context-dependent modeling with $N = 3$) (%)	Quinphones (context-dependent modeling with $N = 5$) (%)
50	94.06	94.29	94.01
100	92.12	93.44	93.56
500	84.46	89.67	90.59
1000	80.34	83.70	88.40

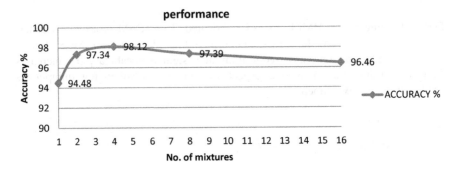

Fig. 3 Performance evaluation of quinphones with varying GMMs

5 Experiments

Experimental section shows the results of different experiments of the work.

5.1 Testing with Varying Vocabulary Sizes

In this section, testing is performed with different modeling units using MFCC with 39 features at front end and with different vocabulary sizes using HMM model at back end. Monophones, triphones, and quinphones that deployed continuous Hindi speech recognition system performances are compared using different vocabulary sizes having HMM model at back end. Table 1 shows performance evaluation of ASR.

5.2 Testing with Varying Gaussian Mixtures

Experiment is performed with Gaussian mixtures varying from 1 to 16; in this, quinphone-based continuous speech recognition gives best results while operating for 4 GMM mixtures. Figure 3 shows its performance evaluation.

6 Conclusion

This paper is based on the use of the wider context phonetics, and here, some examples are shown to describe the use of context-dependent phone models. Quinphone-based acoustic modeling is new in HINDI automatic speech recognition system. In this paper, a comparison between monophones, triphones, and quinphones is implemented by creating large size datasets to check its accuracy. In future, the accuracy of this work can be improved by using different state GMMs.

References

1. O'Shaughnessy, D.: Acoustic analysis for automatic speech recognition. Proceedings of the IEEE vol. 101.5, pp. 1038–1053 (2013)
2. Cai, J., Bouselmi, G., Laprie, Y., Haton, J-P.: Efficient Likelihood Evaluation and Dynamic Gaussian Selection for HMM-Based Speech Recognition. Computer Speech and Language, vol.23, pp. 147–164, (2009)
3. Kumar, A., Dua, M., Choudhary, T.: Continuous Hindi speech recognition using Gaussian mixture HMM, IEEE Student conference on Electrical, Electronics and Computer Science (SCEECS), pp. 1–5, (2014)
4. Becchetti, C., Klucio, P.R. Speech Recognition: Theory and C++ Implementation, 3rd ed., vol. 2, John Wiley & Sons, pp. 121–141, (2008)
5. Cutajar, M., Gatt, E., Grech, I., Casha, O., Micallef, J.: Comparative study of automatic speech recognition techniques. Signal Processing, IET vol. 7.1, pp. 25–46 (2013)
6. Aggarwal, R.k., Dave, M.: Using Gaussian mixtures for Hindi speech recognition system, International Journal of Signal Processing, Image Processing and Pattern Recognition, vol. 4.4, pp. 157–170, (2011)
7. Rybach, D., Riley, M., Alberti, C.: Direct construction of compact context-dependency transducers from data, INTERSPEECH, pp. 218–221, (2010)
8. Schuster, M., Hori, T.: Construction of weighted finite state transducers for very wide context-dependent acoustic models, Automatic Speech Recognition and Understanding, IEEE Workshop, pp. 162–167, (2005)

Slot-Loaded Microstrip Antenna: A Possible Solution for Wide Banding and Attaining Low Cross-Polarization

Ghosh Abhijyoti, Chakraborty Subhradeep, Ghosh Kumar Sanjay, Singh L. Lolit Kumar, Chattopadhyay Sudipta and Basu Banani

Abstract A simple and single element slot-loaded rectangular microstrip antenna has been proposed for broad impedance bandwidth and improved cross-polarized (XP) radiation compared to maximum copolarized (CP) gain without affecting the copolarized radiation pattern. Around 19 dB isolation between CP and XP along with 22% impedance bandwidth is achieved with the proposed structure. The present investigation provides an understanding of simultaneous improvement in impedance bandwidth and the XP radiation characteristics with the present structure.

Keywords Bandwidth · Cross-polarization · Patch reactance · Slot-loaded microstrip antenna

1 Introduction

At the present scenario, most of the wireless communication devices are budding towards tininess and multifunctionality, where rectangular microstrip antenna (RMA) is a good candidate. However, this conventional RMA suffers from some severe disadvantages such as narrow impedance bandwidth and poor polarization purity particularly in its H-plane. The impedance bandwidth of the conventional RMA produces 2–4% bandwidth only [1, 2]. These antennas radiate linearly polarized wave along the broadside of the element called copolarized radiation

G. Abhijyoti · S.L. Lolit Kumar · C. Sudipta (✉)
Department of Electronics and Communication Engineering,
Mizoram University, Aizawl 796004, Mizoram, India
e-mail: sudipta_tutun@yahoo.co.in

C. Subhradeep · G.K. Sanjay
MWT Division, CSIR-CEERI, Pilani 333031, Rajasthan, India

B. Banani
Department of Electronics and Communication Engineering,
NIT-Silchar, Silchar 788010, Assam, India

© Springer Nature Singapore Pte Ltd. 2017
S. Patnaik and F. Popentiu-Vladicescu (eds.), *Recent Developments in Intelligent Computing, Communication and Devices*, Advances in Intelligent Systems and Computing 555, DOI 10.1007/978-981-10-3779-5_14

113

Fig. 1 Schematic representation of the top view of the proposed structure

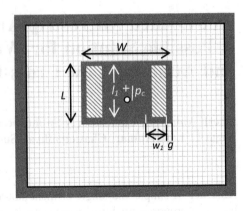

(CP). However, few degree of orthogonally polarized, called cross-polarized (XP), radiation takes place along with the CP radiation. Therefore, such RMA suffers from poor polarization purity (less CP–XP isolation), and it is around 10 dB when operates in X band frequency. This is evidently a restriction to the several wireless applications. Some investigations were reported in [3, 4], where the modifications of conventional patch structure have been employed for the reduction in XP radiation only. Around 16 dB of CP–XP isolation with no improvement in impedance bandwidth is found in those. Two recent investigations [5, 6] show around 23 dB of CP–XP isolation with simple slot and folded-type DGS structure with only 5 and 11% of impedance bandwidth. The employment of shorted patch to reduce XP radiation is a new technique and is reported presently in [7, 8]. More than 25 dB of CP–XP isolation is revealed in [7] with poor impedance bandwidth of only 5%. On the contrary, [8] show some degree of improvement in the bandwidth (around 10.5%; 1.32 GHz at 12.8 GHz) with around 24 dB of CP–XP isolation.

In the present investigation, to take care of both the impedance bandwidth along with XP radiation performance, a simple slot-loaded RMA is proposed and is shown in Fig. 1. It shows around 22% impedance bandwidth with around 19 dB CP–XP isolation. The observed results are justified with the theoretical explanation.

2 Theoretical Background

RMA is an open resonator model where its top and bottom walls are electric walls and four side walls are magnetic walls. Therefore, a slot on the patch surface of a conventional RMA definitely modifies the electromagnetic field beneath the patch and hence alter its input and radiation properties.

A pair of slot with dimension ($l_1 \times w_1$) is cut at patch surface. These are located near nonradiating edges as shown in Fig. 1. The length of the slot (l_1) is chosen in such a way that it resonates at quarter wavelength at dominant mode TM_{10} and half wavelength at second higher order orthogonal mode TM_{02}. The detailed parameters of the proposed antenna are shown in Sect. 3.

Now, each slot $(l_1 \times w_1)$ on the patch is in fact a complementary short quarter-wave dipole. The reactance of short quarter-wave dipole can be given by Balanis [9]

$$X_d = 15\{2Si(kl_1) + \cos(kl_1)[2Si(kl_1) - Si(2kl_1)] \\ - \sin(kl_1)[2Ci(kl_1) - Ci(2kl_1) - Ci(2kw_1^2/l_1)]\} \tag{1}$$

where $k = 2\pi/\lambda_g$, $\lambda_g = \lambda_{10}/\sqrt{\varepsilon_r}$, and $\lambda_{10} = c/(f_r)_{TM_{10}}$

Now, if we increase the width of the slot, it resembles a thick dipole of $l_1/w_1 = 2.5$. The reactance of thick dipole varies slowly with the frequency and hence enhances the bandwidth as is clear from [9].

Now, this reactance X_d comes parallel to patch input impedance Z_p as obtained from [10] as

$$Z_p(f, x_0) = \frac{R_r}{1 + Q_T^2\left(\frac{f}{f_r} - \frac{f_r}{f}\right)^2} - j\left[\frac{R_r Q_T\left(\frac{f}{f_r} - \frac{f_r}{f}\right)}{1 + Q_T^2\left(\frac{f}{f_r} - \frac{f_r}{f}\right)^2}\right] + jX_f \tag{2}$$

where R_r is the resonant resistance of the patch at particular feed position, X_f is feed reactance, and Q_T is the total quality factor corresponding to dominant TM_{10} mode.

The resultant variation of input reactance of the structure with and without slot can be calculated using the above equations and presented in Fig. 2. It reveals that the variation of input reactance varies slowly with frequency when the slots on the patch surface are present compared to conventional RMA without slot on the patch surface. In fact, the sharp variation of input reactance noticeably limits the operating bandwidth of conventional RMA. Therefore, loading of the slots with $l_1/w_1 = 2.5$ is in fact producing the resultant reactance of the whole structure that varies slowly

Fig. 2 Variation of input reactance (with and without slot) versus frequency of the structure

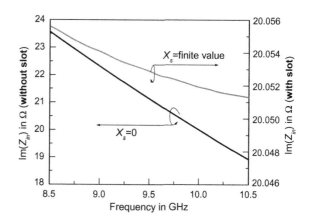

with frequency. Consequently, the impedance bandwidth of the proposed antenna is enhanced compared to conventional RMA. A recent investigation by the present authors [5] clearly demonstrates that the electric fields corresponding to the next higher order orthogonal mode TM_{02}, which in fact is situated near nonradiating edges of the patch, are the major factor which contributes in XP radiation. Therefore, the slots at those regions significantly perturbed the fields of TM_{02} mode and consequently reduce the XP radiation from RMA.

3 Proposed Structure

A RMA is designed using thin copper strip of thickness 0.1 mm, length $L = 8$ mm, and $W = 12$ mm to operate around X band. Taconic's TLY-3-0620 PTFE material ($\varepsilon_r = 2.33$) with thickness $h = 1.58$ mm is utilized as substrate. A pair of slots of length $l_1 = 5$ mm and width $w_1 = 2$ mm are cut at the patch surface and is placed near nonradiating edges at a distance $g = 1$ mm as shown in Fig. 1.

4 Results and Discussions

The parametric studies, using commercially available software package (High-Frequency Structure Simulator; HFSS v. 14) [11], are utilized to investigate the input and radiation characteristics of the proposed structure. The simulation results for input impedance bandwidth (−10 dB) and CP–XP isolation for the proposed antenna for different slot width (w_1) are presented in Table 1. A significant improvement is noted in input impedance bandwidth as well as in CP–XP isolation with the increment of slot width as is clear from the table. The normalized XP pattern of the proposed structure is depicted in Fig. 3. Around 72% of improvement in CP–XP isolation is revealed with $w_1 = 2.0$ mm compared to classical microstrip patch. Further increment of slot width may hamper the dominant mode fields beneath the patch, and hence, we refrain from increasing the width of the slot beyond $w_1 = 2$ mm.

Table 1 Input impedance bandwidth (−10 dB) and CP–XP isolation for the proposed RMA for different width (w_1) of slot at patch surface

Type of structure	w_1 in mm	% Bandwidth	CP–XP isolation in dB
Conventional RMA	No SLOT	5.00	10.00
Proposed slot-loaded RMA structure having different slot widths w_1	0.5	7.00	12.00
	1.0	10.10	14.38
	1.5	19.00	18.00
	2.0	22.38	19.18

Fig. 3 Normalized H-plane XP pattern of the present structure with variable w_1

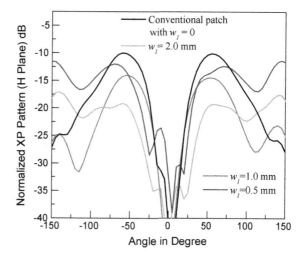

Fig. 4 Reflection coefficient profile for conventional and proposed structure (slot-loaded RMA with defected patch surface)

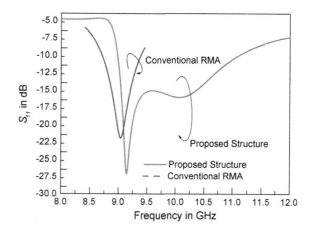

Figure 4 shows the reflection coefficient profile for both conventional and final proposed structure. It is found that both the structures are resonating near 9.1 GHz. It is interesting to note that the introduction of wide slots on the patch surface widens the patch surface significantly. Around 22% of −10 dB impedance bandwidth is revealed from the proposed structure. Figure 5 shows the complete radiation performance of the conventional and proposed structure. Both the figures show that the CP radiation pattern is similar for both the structures in the principal E- and H-planes. Around 4–5 dB suppression in E-plane, XP pattern is evident from proposed structure compared to conventional RMA (Fig. 5a). It is observed from the Fig. 5b that for conventional RMPA, H-plane XP level becomes appreciable (higher than −13 dB) from ±30° to ±80° around the broadside with the peak XP level of −10 dB near ±60°. Therefore, in case of conventional structure,

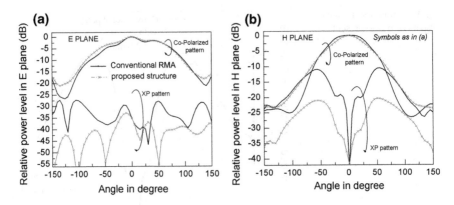

Fig. 5 Radiation pattern for conventional and proposed structures **a** E-plane **b** H-plane

CP–XP isolation is very poor and it is in the order of 10 dB only in its H-plane. A significant improvement is noticed in XP radiation when slots are incorporated on the patch surface. Therefore, the proposed structure results in high CP–XP isolation of more than 19 dB for the present antenna for whole ±150° angular range around broadside. Therefore, the present structure improves both the parameters (input impedance and polarization purity) concurrently and will surely be beneficial for antenna community.

5   Conclusions

Simple single element slot-loaded RMA is proposed for significant improvements in bandwidth and polarization purity. The proposed RMA will surely be helpful for scientific community, researchers, and design engineers looking for small low-profile antenna with stable gain, high polarization purity in wide band of frequencies.

Acknowledgements The authors would like to thank Prof. D. Guha of Institute of Radio Physics and Electronics and Prof. B.N. Basu of Sir J. C. Bose School of Engineering, Mankundu, India, for their support.

Subhradeep Chakraborty and Sanjay Kumar Ghosh thank director, CSIR-CEERI, Pilani, for his permission to publish this work.

References

1. Garg, R., Bhartia, P., Bahl I., and Ittipiboon A.: Microstrip Antenna Design Handbook, Artech House, Norwood, USA, (2001).

2. Guha, D., Antar, Y. M. M. (Eds).: Microstrip and Printed Antennas-New Trends, Techniques and Applications, John Wiley, U.K, (2011).
3. Loffler, D., Wiesbeck, W.: Low-cost X-polarised broadband PCS antenna with low cross-polarisation. Electronics Letters, 35, 1689–1691, (1999).
4. Granholm, J., Woelders, Kim.: Dual Polarization Stacked Microstrip Patch Antenna Array With Very Low Cross-Polarization. IEEE Transactions on Antennas And Propagation, 49, 1393–1402, (2001).
5. Ghosh, A., Ghosh, D., Chattopadhyay, S., Singh, L. L. K.: Rectangular Microstrip Antenna on Slot Type Defected Ground for Reduced Cross Polarized Radiation. IEEE Antennas and Wireless Propagation Letters, 14, 324–328, (2015).
6. Kumar, C., Guha, D.: DGS integrated Rectangular microstrip patch for improved polarization purity with wide impedance bandwidth. IET Micro. Ant. & Prop., 8, 589–596, (2014).
7. Ghosh, D. Ghosh, S. K., Chattopadhyay, S., Nandi, S., Chakraborty, D., Anand, R., Raj, R., Ghosh, A.: Physical and Quantitative Analysis of Compact Rectangular Microstrip Antenna with Shorted Non-Radiating Edges for Reduced Cross-Polarized Radiation Using Modified Cavity Model. IEEE Antennas and Propagation Magazine, 56, 61–72, (2014).
8. Poddar, R., Chakraborty, S. and Chattopadhyay, S.: Improved Cross Polarization and Broad Impedance Bandwidth from Simple Single Element Shorted Rectangular Microstrip Patch: Theory and Experiment, Frequenz, 70, 1–9, (2016).
9. Balanis, C.A.: Antenna Theory: Analysis and Design, 2nd Ed., Wiley, USA, (2001).
10. Chattopadhyay, S. et. al.: Input Impedance of Probe fed Rectangular Microstrip Antenna with Air gap and Aspect Ratio. IET Micro. Ant. & Prop., 3, 1151–1156, (2009).
11. HFSS, "High Frequency Structure Simulator, Version 14," Ansoft Corp.

Fractal PKC-Based Key Management Scheme for Wireless Sensor Networks

Shantala Devi Patil, B.P. Vijayakumar and Kiran Kumari Patil

Abstract We intend to introduce through this paper a new fractal-based crypto-graphic method to the world of wireless sensor networks (WSNs). Fractal cryptography is part of chaos cryptography with strong security features due to its chaotic nature and complex structure, making it more preferable than the prevailing cryptographic schemes. In this paper, a novel fractal PKC-based key management scheme for heterogeneous mobile WSNs is proposed. The scheme uses the Mandelbrot and Julia fractal sets due to their strong interconnectivity and is more efficient for resource-constrained WSNs with short key size than RSA and ECC. We compare the efficiency of the proposed scheme with the existing schemes.

Keywords Wireless sensor networks · Key management · Fractal PKC

1 Introduction

Wireless sensor network (WSN) is a special type of ad hoc network made up of many small sensor nodes, widely applicable to the real world such as military, physiological and ecological scenarios. Being used in critical applications, the communication among the nodes must be secure but the wireless nature makes it

S.D. Patil (✉)
Computer Science and Engineering, REVA Institute of Technology and Management,
Bangalore, India
e-mail: shantaladevipatil@gmail.com

B.P. Vijayakumar
Information Science and Engineering, M S Ramaiah Institute of Technology,
Bangalore, India
e-mail: vijaykbp@yahoo.co.in

K.K. Patil
School of Computing and Information Technology, REVA University,
Bangalore, India
e-mail: kiran_b_patil@rediffmail.com

© Springer Nature Singapore Pte Ltd. 2017
S. Patnaik and F. Popentiu-Vladicescu (eds.), *Recent Developments in Intelligent Computing, Communication and Devices*, Advances in Intelligent Systems and Computing 555, DOI 10.1007/978-981-10-3779-5_15

vulnerable to attacks. WSN also faces resource limitation with respect to processing, storage, transmission range, bandwidth and short-lived battery. Securing such a constrained WSN is more challenging. The key management schemes developed must be able to address the shortcomings in sensor networks. In this paper, we propose a new fractal PKC-based key management scheme that is able to generate strong keys and securely distribute these keys along with support for node mobility. We use Mandelbrot set to generate the public–private key pair and Julia set to exchange a shared private key.

The rest of the paper is organised as follows: in Sect. 2, we bring out the shortcomings of the various previously published key management schemes. In Sect. 3, we discuss the preliminaries required to propose our scheme. In Sect. 4, we discuss our fractal PKC-based key management scheme (FPKM). In Sect. 5, we evaluate the performance of the FPKM and conclude in Sect. 6.

2 Related Work

The key management schemes proposed in the literature are based on symmetric key (SKC) or asymmetric key (PKC) techniques, with the first one presumed to be the right choice for WSN. The SKC schemes [1, 2] focused on static and homogeneous scenario but are a poor choice for heterogeneous mobile WSN as they require more memory space, high communication overhead, no mobility support and are not scalable. It was presumed that PKC schemes were infeasible which was proved otherwise in [3]. PKC schemes achieve secure key management with node authentication, are resilient to attacks over nodes and also scalable. For mobile WSN, many key management schemes are proposed [4–10] based on PKC such as RSA algorithm, elliptic curve cryptography algorithm ECC or identity based algorithm. References [5, 7] are key management schemes based on certificate exchange to establish keys. These schemes incur high communication and computation overhead with base station performing the certificate management causing a bottleneck and hence infeasible WSN. Schemes in [6] are based on Diffie–Hellman PKC for mobile WSN, infeasible due to large key sizes and high communication overhead. In [4], identity—PKC schemes are proposed to overcome the shortcomings of previous two schemes, but suffer computationally due to expensive pairing operations. To the best of our knowledge, there are no efficient key management schemes that are strongly secure, lightweight and mobility support for heterogeneous WSNs.

Fractals are part of chaos systems that are created iteratively from similar smaller geometric components. In [11], author proposed a key management protocol based on these fractals and compared with the Diffie–Hellman protocol. This study proves that the fractal-based public key cryptography is more efficient than the traditional public key cryptographic techniques. In this paper, we extend the application of fractal-based public key cryptography proposed in [11] to the WSNs and propose an effective key management scheme for WSNs called FPKM. This proposed scheme eliminates the need for certificates, large key sizes and heavy computations.

3 Preliminaries

3.1 System Model

The clustered heterogeneous WSN scenario as described in [12] is considered. The nodes are deployed randomly in the destined area and are of two energy levels such as high-energy nodes 'HN' and low-energy nodes 'LN'. The HN are less in number but with high resource reserve, making preferable candidates for the role of the cluster heads 'CH', but LN are in huge number with low energy resource and these perform menial tasks of the network. The member nodes 'CM' of the cluster 'CL' move from one cluster to another. As they move, they register themselves to different CH within their vicinity. The CH control the functioning of the cluster and is responsible for managing node mobility and periodically reporting status to the BS. The BS along with managing network is also responsible for initial system set-up, key generation and keying material assignment to the nodes of the sensor network.

The various keys required for communication are *public–private key pair (Pub, Pri)*: each node is assigned with a public–private key pair that is pre-deployed by the base station, *cluster key (CK)*: this key is used to securely broadcast all the cluster members, *master key (MK)*: is used to securely broadcast with the cluster heads in the network and *pairwise secret key (K)*: each pair of nodes share a secret key k for unicast communication.

3.2 Fractal-based Key Pair Generation and Key Exchange

For our work, we use Mandelbrot and Julia fractal sets for generation of public–private key pair and exchange of secret keys among nodes. To illustrate the key generation, say two nodes Alice and Bob want to communicate. Let c be globally known.

Step 1 Alice selects e and n, generates a public key $Z_n e$ and private key (e,n) pair by using $Z_n e = Z_{n-1} \times c^2 \times e$; $Z,e,c \in C$; $n \in Z$.

Step 2 Bob selects k and d, generates a public key $Z_k d$ and private key (d,k) pair. Using $Z_k d = Z_{k-1} \times c^2 \times d$; $Z,d,c \in C$; $k \in Z$.

Step 3 Alice and Bob broadcast the public key.

Step 4 Alice calculates the secret key using the Julia fractal set based on the parameters and Bob's public key, using equation $K = C^{n-x} \times (Z_k d)_n e$.

Step 5 Bob calculates the secret key using the Julia fractal set based on the parameters and Alice's public key, using $K = C^{k-x} \times (Z_n e)_k d$.

Step 6 The secret key produced by Alice and Bob will be the same $K = C^{k-x} \times (Z_n e)_k d = C^{n-x} \times (Z_k d)_n e$.

4 Fractal PKC-based Key Management Scheme: FPKM

The proposed scheme is having 3 phases: (1) set-up phase, (2) clustering phase and (3) rekeying phase.

4.1 Set-up Phase

In this phase, the BS assigns each node a unique identifier NID, generates keys, keying material, individual secret key SKN and pre-deploys in each of the node.

- *Generation of public–private key pair {PUB_N, PRI_N}*: the generation of key pair for each node is based on the fractal key generation method as discussed in Sect. 3.2. For each node with identity IDN, the BS selects global variable '*c*' and $PRI_N = (e,n)$. Using Mandelbrot fractal function in step 1, $PUB_N = Z_n e$ and $PRI_N = (e,n)$ are calculated.
- *Generation of individual secret key SK_N*: these keys are used to alert the BS in the case of any critical events. For each node with identity ID_N, a secret individual key is calculated using simple hashing $SK_N = H(RN_N \| N_{ID})$, where random number RN ϵ Zq.

Each node is then preloaded with the following sequence before the deployment {N_{ID}, (PRI_N, PUB_N), S_{KN}, PUB_{BS}, R_N, C}. The BS initially creates three lists: valid nodes VN = {IDN, (PRI_N, PUB_N), S_{KN}, R_N}, compromised nodes CN = {ϕ} initially set to null and the CH nodes CHN = {CH_{ID}, CH_{LOC}, CH_{ENE}, PUB_{CH}, RN_{CH}}.

4.2 Clustering Phase

In this phase, CH register the nodes within its communication range as CM, establishes pairwise key SK {CH_{ID}, N_{ID}} using the fractal key exchange as described in Sect. 3.2 and a cluster key CK. Similarly, the generation of master key MK and the pairwise keys for the BS communicates with the CH.

- *Node registration*: the CH broadcast join message JOIN_MSG to all the nodes within its communication range. The nodes receiving this check the details from each CH and revert back with the register message REG_MSG to the closest CH with high energy after lapse of '*t1*' threshold time. The CH and CM store the exchanged details.

$$JOIN\_MSG = \{CH_{ID}, CH_{LOC}, CH_{ENE}, PUB_{CH}\}; REG\_MSG = \{N_{ID}, PUB_N\}$$

- *Mutual registration of orphan nodes*: if any node finds itself unaffiliated to any cluster even after elapse of *t2* the threshold time for node registration, then we term such a node as 'orphan. Such nodes broadcast MUTUAL_REG message to nodes in its radio range, and the nodes receiving this respond back with a MUTUAL_ACK message. The orphan node may receive many such messages, check the details from each CH and revert back with the register message REG_MSG to the closest CH with high energy.

$$MUTUAL\_REG = \{N_{ID1}, PUB_N\}; MUTUAL\_ACK$$
$$= \{CH_{ID}, CH_{LOC}, CH_{ENE}, PUB_{CH}, N_{ID2}\};$$

- *Pairwise key establishment between CH and CM;$K_{ID} = SK\{CH_{ID}, N_{ID}\}$*: the CH and CM both establish $SK\{CH_{ID}, N_{ID}\}$ using Julia fractal function with the details exchanged during node registration. The CM uses to its PUB_N, PRI_{CH} and c to deduce the $KID = SK\{CH_{ID}, N_{ID}\} \Rightarrow$ same $C^{k-x} \times (Z_n e)_k d = C^{n-x} \times (Z_k d)_n e$.
- *Cluster key generation*: CH initiate this cluster key generation process, using binary logical key tree. The leaf nodes in the tree are the pairwise key between the CH and CM established using fractal key exchange. Let us assume that cluster head has 8 cluster members and the pairwise keys shared with these CM are $K1, K2, \dots K8$ as shown in Fig. 1.
- *Pairwise key establishment between CH_{ID} and $BS = SK\{CH_{ID}, BS\}$*: the CH and BS both establish $SK\{CH_{ID}, BS\}$ using Julia fractal function as described in Sect. 3.2.
- *Master key generation MK*: the MK is generated by the base station in a similar way as that of the CK using exclusive pairwise keys shared between the CH and the BS as shown in Fig. 1.

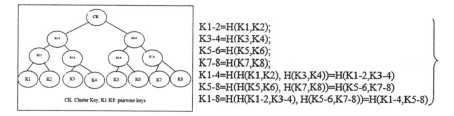

K1-2=H(K1,K2);
K3-4=H(K3,K4);
K5-6=H(K5,K6);
K7-8=H(K7,K8);
K1-4=H(H(K1,K2), H(K3,K4))=H(K1-2,K3-4)
K5-8=H(H(K5,K6), H(K7,K8))=H(K5-6,K7-8)
K1-8=H(H(K1-2,K3-4), H(K5-6,K7-8))=H(K1-4,K5-8)

Fig. 1 Cluster key generation

4.3 Rekeying Phase

Rekeying in mobile WSNs is performed when the CH move in and out of network, when the CM join and leave the cluster, when new members join the network at arbitrary time and for long-lived networks.

- Key updation:

    - *New CH join network*: let us consider the key updation case of a CH joining a network and the same can be extended for the other cases. The new CH node sends a REG-CH_MSG to the BS, and the BS checks the VN list to validate the requesting CH. On validation, the BS generates pairwise key with new CH. The BS also updates the MK as follows: the BS unicasts the encrypts MKnew1 with the pairwise key of the new CH and unicasts it to the new CH. Also, the BS encrypts the MKnew1 with previous MK and broadcasts to the other CH. The other CHs decryp the message to obtain the MKnew1.
    - *Node entry into a cluster or new node joins the network*: when a node enters the CL, it uses the mutual registration method to register with the new CH. The same can be extended to a new node joining the network. Once registered, the CH takes lead to generate pairwise keys and update the CK.

- Key Revocation:
  Key revocation is performed on nodes that are compromised or have left the network or have joined some new cluster or dead due to exhaustion of resources.

    - *CH leaving the network* and the same procedure can be extended to the other scenarios. When the BS learns that a cluster head is either compromised or dead from the other neighbouring nodes, the BS creates a MKnew2, encrypts this with the pairwise keys exclusively shared between the BS and the CH and then communicates to the other CH along with the revoked CH_{ID}. The CH decrypt this message from the BS, updates the MK to MKnew2 and the C_N list of the CH.
    - *Node exit from a cluster*: when a CM move out of a cluster, the new CH to whom this node registers will broadcast the credentials of this node and the previous CH updated the list of nodes in the CL.

5 The Performance Evaluation of Proposed Scheme

Some of the prominent public key algorithms are RSA, ECC, etc. For the RSA public key generation algorithm for a key size of 1024-bits, the time taken for keying is 22 s. The schemes based on ECC use 160-bit keys and time taken for keying is 1.62 s. In [11], fractal-based PKC uses 128-bit keys and requires less than 250 ms for keying. The key management scheme fractal-based PKC provides better security due to chaotic nature. Such schemes have better trade-off between key size and time taken for keying, when compared to RSA- and ECC-based schemes.

The proposed + scheme is implemented using MATLAB with 100 nodes randomly deployed. We assume only 10% of total nodes are nodes with high energy and rest of the nodes are with low energy. We evaluate the performance of the FPKM against ECHCKM and Khamy's scheme.

Key storage space: the number of keys stored by the BS for 'm' nodes of the network is $(2m + 4)$; for n nodes in the cluster, the number of keys stored is $(n + 5)$ and the number of keys stored on individual network is 5 keys, whereas in the ECHCKM scheme, the number of keys stored by the BS is $(2m + 4)$ keys, in each cluster head is $(2n + 7)$ and in each node is 6. Similarly, we calculate the total number of keys for Khamy's scheme which is $10mn + 8m + 4$. From above calculations, we can infer that the total number of keys for FPKM is significantly less than other two schemes.

Number of message exchanges to establish keys: in FPKM scheme, the number of message exchanges to establish keys is also considerably less than ECHCKM and Khamy's scheme.

Time consumption for key establishment: in the proposed FPKM scheme, there is no need to explicitly establish the pairwise keys, and it is implicitly carried out during the node registration to the CH. The time taken for key establishment is equal to the time taken by each node to register itself to the CH. In our proposed FPKM scheme, we assume the number of CH to be fixed at any point of time. Let TimeKE be the time consumption of key establishment in the network. This can be obtained by calculating the TimeKECi, time taken for key establishment by each cluster Ci in the network. The TimeKECi is influenced by the number of CM in Ci. More the CM, more is the time taken for key establishment in Ci. Hence, the time taken for establishing the keys is the highest time consumed by any cluster Ci to complete the process.

6 Conclusion

The proposed scheme is first of its kind with fractal PKC applied to resource-constrained WSN. In this paper, we present a new fractal PKC-based key management scheme (FPKM) for WSNs with heterogeneity and mobility. Compared to the existing schemes based on RSA, ECC, and ID-based PKC, our scheme provides resiliency with low storage, computational and communication overhead. From performance analysis, it is evident that fractal PKC uses short key size and also is very secure due to complex chaotic nature of Mandelbrot and Julian fractals.

References

1. Chan H., Perrig A., Song D.: Random key pre-distribution schemes for sensor networks. In: Proc. IEEE Symp. SP, pp. 197–213, (May 2003)
2. Du W., Deng J., Y. S. Han., P. K. Varshney.: A key pre-distribution scheme for sensor networks using deployment knowledge. In: IEEE Trans Dependable Secure Comput., vol. 3, no. 1, pp. 62–77, (2006)
3. N. Gura., A. Patel., A. Wander., H. Eberle., S. C. Shantz.: Comparing elliptic curve cryptography and RSA on 8-bit CPUs. In: Proc. 6th Int. Workshop Cryptograph. Hardw. Embedded Syst., pp. 119–132, (2004)
4. Chatterjee K., De., Gupta D.: An improved ID-based key management scheme in wireless sensor network. In: Proc. 3rd Int. Conf. ICSI, vol. 7332, pp. 351–359, (2012)
5. Alagheband M R., Aref M R.: Dynamic and secure key management model for hierarchical heterogeneous sensor networks. In: IET Inf. Secur., vol. 6, no. 4, pp. 271–280, (2012)
6. Chuang I H., Su W T., Wu C Y., Hsu J P., Y.-H. Kuo.: Two layered dynamic key management in mobile and long-lived cluster based wireless sensor networks. In: Proc. IEEE WCNC, pp. 4145–4150, (2007)
7. Zhang X., He J., Wei Q.: EDDK: Energy-efficient distributed deterministic key management for wireless sensor networks. In: EURASIP J. Wireless Comm. Netw. vol. 2011, pp. 1–11, (2011)
8. Szczechowiak P., Oliveira L B., Scott M., Collier M., Dahab R.: NanoECC: Testing the limits of elliptic curve cryptography in sensor networks. In: Proc. 5th Eur. Conf. WSN, vol. 4913, pp. 305–320, (2008)
9. Hamed A., EL-Khamy E.: New Low Complexity Key Exchange and Encryption protocols for Wireless Sensor Networks Clusters based on Elliptic Curve Cryptography. In: 26th NATIONAL RADIO SCIENCE CONFERENCE, NRSC, pp. 1–13, (2009)
10. Srikanta Kumar Sahoo., Manmanth Narayan Sahoo.: An Elliptic Curve based Hierarchical Cluster Key Management in WSN. In: ICACNI 209
11. M Alia., A Samsudin.: New Key Exchange Protocol Based on Mandelbrot and Julia Fractal Set. In: International Journal of Computer Science and Network Security, 7 pp. 302–307, (2007)
12. Patil S D., Vijayakumar B P.: Clustering in Mobile Wireless Sensor Networks: A Review. In: International Conference on Innovations in computing and networking, (2016)

Histogram-Based Human Segmentation Technique for Infrared Images

Di Wu, Zuofeng Zhou, Hongtao Yang and Jianzhong Cao

Abstract Human detection in infrared video surveillance system is a challenging issue of computer vision. Effective human segmentation plays an important role in human detection. However, occlusion between different people makes it difficult to segment human groups. In this paper, we propose a new method for infrared human segmentation based on the histogram information. After selecting regions of interest with background subtraction, each connected human region is separated into single ones by analyzing histogram trend and calculating peak number. Experiment results show the accuracy of our method.

Keywords Histogram · Human segmentation · Infrared images

1 Introduction

Generally, infrared human detection algorithm comprises two stages, human candidate regions detection and human candidate regions recognition. The fundamental step for human candidate regions detection is human segmentation, as shown in Fig. 1. Many methods have been proposed to segment infrared images, such as template matching [1], thresholding [2], and region growing [3], but most of them are too complex to operate and invalid for occlusion problems.

The rest of this paper is organized as follows: Sect. 2 describes the proposed human segmentation technique. Section 3 describes the experiment results and discussion of the proposed technique. Section 4 presents the conclusion and the future work.

D. Wu (✉) · Z. Zhou · H. Yang · J. Cao
Xi'an Institute of Optics and Precision Mechanics of CAS,
Xi'an 710119, People's Republic of China
e-mail: wudi@opt.cn

D. Wu
University of Chinese Academy of Sciences, Beijing 100049,
People's Republic of China

Fig. 1 General mechanism of human candidate region detection

Fig. 2 Flowchart of selection of ROIs

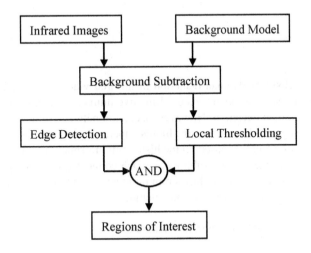

2 The Proposed Method

We have concluded from [4–6] that human regions are usually brighter than their surroundings in infrared images and human bodies have more vertical edges than other disturbing objects. Besides, the histogram of single human region appears similar trend and only one peak for both directions. In the proposed method, we make a combination of edge detector and brightness operator to extract foreground regions. Then, the histogram information of each connected region is analyzed to eliminate false regions and separate each human candidate region into single ones.

2.1 Selection of ROIs

Regions of interest (ROIs) imply those regions which possibly include humans. Additional background frames is required to establish background models. Foreground regions containing humans will be obtained as shown in Fig. 2.

2.2 Human Candidates Segmentation

In order to remove non-human objects in ROIs, each connected region is analyzed by means of histogram evaluation. If the trend of a histogram is monotonic, the corresponding region is too uniform to represent a human. Besides, it is very tough to identify a human in infrared images when many people are walking together closely or interact with each other frequently. It is necessary to segment overlapped human regions to facilitate further recognition process.

The algorithm works as the following steps:

1. Scanning each connected region by columns, adding the gray-level value belonging to each column pixel to obtain histograms.
2. Eliminating corresponding connected regions if the histogram trend is monotonic.
3. Calculating the peak number of each remaining connected region, if there is only one peak in a histogram, it goes to step 4; otherwise, the related connected region will be segmented according to the valley of the histogram.
4. Scanning each connected region by rows, adding the gray-level value belonging to each row pixel to obtain histograms.
5. Calculating the peak number of each remaining connected region, if there is more than one peak in a histogram, the related connected region will be segmented according to the valley of the histogram.

This stage aims to make sure that there is only one human candidate in each connected region, since samples of recognition process are always single humans.

3 Experiment Results

The proposed histogram-based human segmentation technique has been tested on a series of infrared images from OSU thermal pedestrian database. The following table shows the experiment results of applying our technique to detect humans in infrared images, where #TP, #FP, and #People represent the number of true positive, false positive and humans, respectively. For analyzing the accuracy of the different human segmentation techniques, we use parameter sensitivity. The sensitivity reports humans which are correctly identified by the algorithm, where a high-sensitivity value means a high detection rate of humans. When compared with method described in [7], our technique operates better on sensitivity performance obviously, as shown in Table 1.

Table 1 Detection results for OSU thermal pedestrian database

Sequence No.	#Frames	#People	#TP		#FP		Sensitivity	
			[7]	Ours	[7]	Ours	[7]	Ours
1	31	91	88	91	0	0	0.97	1.00
2	28	100	94	98	0	0	0.94	0.98
3	23	101	101	101	1	0	1.00	1.00
4	18	109	107	108	1	1	0.98	0.99
5	23	101	90	101	0	0	0.89	1.00
6	18	97	93	96	0	0	0.96	0.99
7	22	94	92	93	0	0	0.98	0.99
8	24	99	75	92	1	1	0.76	0.93
9	73	96	95	96	0	0	1.00	1.00
10	24	97	95	97	3	1	0.98	1.00
1–10	284	984	930	973	6	3	0.95	0.99

4 Conclusion

In this paper, we present a histogram-based human segmentation technique for infrared images. The proposed method is working on the principal that the histogram of single human region satisfies specific constraint. Experiment results show that a fairly high detection rate of humans can be achieved with the proposed method. In the future work, we plan on extending the database to infrared videos captured by our own infrared camera, possibly including more disturbing objects in the scene.

References

1. Nanda, H., Davis, L.: Probabilistic template based pedestrian detection in infrared videos. Proceedings of the IEEE Intelligent Vehicle Symposium, vol. 1, pp. 15–20 (2002)
2. Bertozzi, M., Broggi, A., Caraffi, C., Rose, M.D., Felisa, M., Vezzoni, G.: Pedestrian detection by means of far-infrared stereo vision. Computer Vision and Image Understanding. vol. 106, no. 2, pp. 194–204 (2007)
3. Bingjie, Z., Wei, G., Zongxi, S.: Infrared and visible fusion based on region growing and contour let transform. In: International Symposium on Photoelectronic Detection and Imaging, vol. 8907, Proc. SPIE. Beijing (2013)
4. Jeon, E.S., Choi, J., Lee, J.H., Shin, K.Y., Kim, Y.G., Le, T.T., Park, K.R.: Human detection based on the generation of a background image by using a far-infrared light camera, Sensors, vol. 15, no. 3, pp. 6763–6788 (2015)
5. Wang, J.T., Chen, D., Chen, H., Yang, J.: On pedestrian detection nad tracking in infrared videos, Pattern Recognition Letters, vol. 33, no. 6, pp. 775–785 (2012)
6. Castillo, J.C., Cuerda, J.S., Caballero, A.F.: Robust people segmentation by static infrared surveillance camera, Lecture Notes in Computer Science 6096, pp. 348–357 (2010)
7. Davis, J.W., Keck, M.A.: A two-stage template approach to person detection in thermal imagery, Proceedings of the Seventh IEEE Workshop on Application of Computer Vision, vol. 1, pp. 364–369 (2005)

Medical Image Segmentation Based on Beta Mixture Distribution for Effective Identification of Lesions

S. Anuradha and C.H. Satyanarayana

Abstract Brain imaging plays a vital role toward the identification of diseases such as seizures, lesions, sclerosis, and the other inhomogeneities. Methodologies for effective and efficient regulation of these diseases are to be planned so as to overcome the issues of mortality. This chapter highlights the contributions using beta mixture models in this direction. The experimentation is carried out in a MATLAB environment and the results are tabulated based on BRAINWEB images. The results are also compared with those of the existing models based on GMM using the performance evaluation parameters such as average difference, maximum distance, and image fidelity. The results showcase that the proposed methodology overcomes the GMM in all respects and it gives good recognition accuracy. The developed model can also be used for identifying the other diseases of the brain.

Keywords Beta mixture model · Mortality · Sclerosis · Inhomogeneity · Lesion · Mixture model

1 Introduction

Among the human organs, brain is considered to be vital, and therefore, it is customary to safeguard the brain from diseases. Most of the diseases that are majorly coined include tumor, sclerosis, lesion, and the other inhomogeneities. Tumors are caused due to rupture among the nervous system, excessive blood flow, diseases, or even due to the injuries. Therefore, identifying the root cause is a major concern whereby effective planning of the treatment is to be carried out. Among the other diseases, lesions play a vital importance. The individual suffering with the lesion may be

S. Anuradha (✉)
Department of CSE, GIT Gitam University, Visakhapatnam, India
e-mail: ganuharsh@gmail.com

C.H. Satyanarayana
Department of CSE, JNTU, Kakinada, India
e-mail: chsatyanarayana@yahoo.com

© Springer Nature Singapore Pte Ltd. 2017
S. Patnaik and F. Popentiu-Vladicescu (eds.), *Recent Developments in Intelligent Computing, Communication and Devices*, Advances in Intelligent Systems and Computing 555, DOI 10.1007/978-981-10-3779-5_17

witnessed with symptoms such as dizziness, changes in blood pressure levels, sweating, vomiting, and even faint. There may be several other diseases which share similar symptoms. Therefore, the challenge associated is to identify the actual disease, so that the treatment can be planned accordingly. However, the models available in the literature consider very few factors and the classification and identification of the disease are subjected to mere consideration of a typical symptom [1–4]. These models are considered to be degenerative, where the patterns are the attributes of the pixels related to the diseased person who are not taken into consideration [5–7]. Other methodologies also highlighted in the literature are based on the pseudo-supervised models and generative model-based approaches. Among these techniques, generative models are considered to be more effective because of the reason that they consider the parameters inside the diseased images before the identification of the disease [8–10]. Among the generative models, approaches based on normal distribution are highly focused. This is because of the assumption that the anatomy of the diseased image is normal in shape. However, whenever a deformity takes place, the normality of the distribution gets effected and thereby resulting into asymmetric distribution [11, 12]. With this assumption, researchers have therefore considered asymmetric models such as gamma, lognormal for the identification of the diseases. However, most of the image regions within the damaged image may contain pixels where the intensity of the damaged pixels is low compared to that of non-damaged pixels. Therefore to model such combination of frequencies, one needs to consider a model which is partly symmetric and partly asymmetric. Here in this chapter, a proposal is made by considering beta mixture models. The initial parameters of the distribution are updated using the EM algorithm, and based on these updated parameters, the segmentation algorithm is carried out in a maximum posteriori approach. The rest of the chapter is organized as follows: In Sect. 2, the probability distribution functions of the beta mixture model together with the updated parameters using the EM algorithm are presented. The corresponding Sect. 3 of this chapter highlights the clustering algorithm based on fuzzy c-means. The dataset considered is highlighted in Sect. 4, and segmentation processes together with the experimentation are presented in Sect. 5 of this chapter. Section 6 highlights the results derived together with the performance analysis. Section 7 concludes with the summarization together with the scope for further research.

2 Probability Distribution Functions of Beta Mixture Model Using the EM Algorithm

The probability density function of the beta mixture model is given by

$$B(x_i, \alpha_i, \beta_i) = \sum_{i=1}^{L} \pi_i f_i(\mu_i | \alpha_i, \beta_i) \tag{1}$$

The basic parameters associated are to be significantly processed, for which EM algorithm is considered. The initial estimates of the model are acquired by using k-means algorithm. The parameters are updated by using the EM algorithm. The updated equations for the parameters (π, μ, σ) are given as follows:

$$\alpha_k^{l+1} = \frac{1}{N} \sum_{s=1}^{N} \frac{\alpha_k^{(l)} g_i\left(x_i, \theta^{(l)}\right)}{h\left(x_s, \theta^{(l)}\right)} \tag{2}$$

$$\beta_k^{i+1} = \frac{1}{\alpha_k} \sum_{s=1}^{N} \left[t_k\left(x_i, \theta^{(l)}\right) \right] \tag{3}$$

$$\mu_k^{i+1} = \left[x^{\alpha_i x - 1}(1 - x)^{\beta_i x - 1} \right] \cdot e^{\left(\frac{x-\mu}{\sigma}\right)^2} \left[\beta\left(x_{i\mu}, x_{\mu i}\right)^{-1} \right] \tag{4}$$

In this mixture model, it has twofold advantage such that it can portray the behavior of symmetric models as well as asymmetric models. Therefore, the usage of this model helps to validate all those images where the peaks are low, i.e., of the image having low intensity values and it can also be effective in cases of high peaks. The updated parameters formulate the basis for the identification of the diseases more appropriately.

3 Clustering Algorithm Based on Fuzzy c-Means

The medical data are mostly unstructured in nature, to convert into meaningful data, one need to use clustering algorithms. However, among the available clustering algorithms, fuzzy c-means is preferred because of its ability in identifying the partly damaged cells. The step-by-step process is depicted below:

Step 1: Take an image from the dataset and preprocess.
Step 2: Cluster the image into regions using fuzzy c-means algorithm.
Step 3: Update the initial parameters of the beta mixture model.
Step 4: For each image pixel, find the corresponding probability density function.
Step 5: Consider the log-likelihood estimate and group the pixels into appropriate image regions.

Fuzzy c-means algorithm:

Step 1: Consider an arbitrary value for partitioning the data, and let it be W_{ij} where i denotes pixel and j denotes cluster.
Step 2: Compute the process for the other pixels and clusters.
Step 3: Estimate the centroid using the following formula:

$$SSE(C_1, C_2, \ldots, C_k) = \sum_{j=1}^{k} \sum_{i=1}^{m} w_{ij}^{p} \text{dist}(x_i, c_j)^2 \tag{5}$$

c_j denotes clusters.
p number of partitions.

Step 4: Recompute the partition W_{ij} using the following formula:

$$W_{ij} = \frac{1/\text{dist}(x_i, c_j)^2}{\sum_{q=1}^{k} 1/\text{dist}(x_i, c_j)^2} \tag{6}$$

Step 5: Repeat the process until the centroids converge.

4 Dataset

In order to present the proposed methodology, a benchmark dataset consisting of brain images is taken from the UCI dataset. This dataset consists of the various diseases of the brain, and for the training purpose, we have considered 2000 samples and for testing 100 sample images.

5 Segmentation Algorithm

The stepwise process of segmentation is underlined below:

Step 1: Consider the images from the dataset to process.
Step 2: Apply the clustering algorithm so as to formulate the image regions.
Step 3: Consider each pixel from the image regions and apply the probability density function of the beta mixture model by integrating the updated parameters.
Step 4: Take an individual pixel from the actual image and substitute it in the probability density function.
Step 5: Identify the likelihood of the pixel with a particular probability density function falling into a particular region by considering the log-likelihood estimate.
Step 6: Regroup the component likelihood function for the reconstruction.
Step 7: Evaluate the performance using metrics such as average difference, maximum distance, and image fidelity.

5.1 Experimentation

In order to propose the developed model, the experimentation is conducted in a MATLAB environment by considering 75 test images and 25 training images. Each of the images is normalized and preprocessed for noise elimination. The pixels within the image are considered, and the segmentation algorithm is applied to formulate the regions. Each of the region distributions is identified by using the probability density function of the mixture model considered. The step-by-step approach of the segmentation process is carried out as underlined in the segmentation algorithm proposed in Sect. 5. In order to evaluate the model, the performance metrics such as average difference, maximum distance, and image fidelity are considered, and the results are compared to those of the existing model based on GMM.

6   Results Derived and Performance Analysis

See Table 1 and Figs. 1, 2, and 3.

Table 1 Performance evaluation

Image	Quality metric	Finite beta GMM with fuzzy c-means (FBGMM + FCM)	GMM
BOS1	Average difference	0.81	0.819
	Maximum distance	0.88	0.489
	Image fidelity	0.88	0.38
BOS2	Average difference	0.90	0.48
	Maximum distance	0.81	0.19
	Image fidelity	0.88	0.364
BOS3	Average difference	0.83	0.68
	Maximum distance	0.86	0.48
	Image fidelity	0.88	0.63

(continued)

Table 1 (continued)

Image	Quality metric	Finite beta GMM with fuzzy c-means (FBGMM + FCM)	GMM
B0S4	Average difference	0.61	0.39
	Maximum distance	0.93	0.33
	Image fidelity	0.81	0.38
B0S5	Average difference	0.83	0.34
	Maximum distance	0.89	0.29
	Image fidelity	0.88	0.48

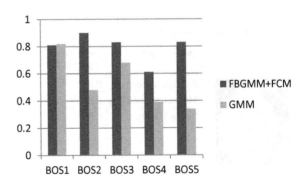

Fig. 1 Comparison of FBGMM-FCM using average difference

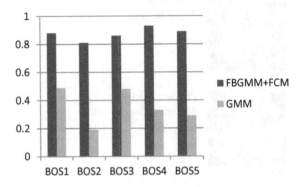

Fig. 2 Comparison of FBGMM-FCM using maximum distance

Fig. 3 Comparison of FBGMM-FCM using image fidelity

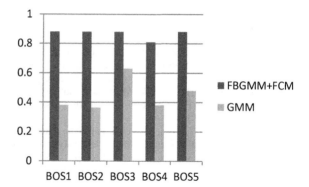

7 Conclusion

A methodology for identification of a medical deformity based on the lesions is presented in this chapter. The parameters at the diseased region are identified and are modeled using beta mixture model and the results derived are experimented on real-time benchmark images. The experimentation is carried out using MATLAB and the results derived thereof are validated using the quality metrics such as average difference, maximum distance, and image fidelity. The outputs are also compared with those of the existing model based on GMM, and the results are presented in Table 1. From Table 1, it is shown that the average difference of the constructed model is performing far better than the model based on GMM. The results are the same in case of all the images and for all the metrics. The methodology developed can be very much useful for the identification of the other diseases that are prone to human brain.

References

1. Miller DH, Albert PS, Barkhof F, et al.: Guidelines for the use of magnetic resonance techniques in monitoring the treatment of multiple sclerosis. *Ann Neurol*; 39:6–16 (1996).
2. Paresh Chandra Barman et al.: MRI Image segmentation using level set method and implement an medical diagnosis system *Computer Science & Engineering: An International Journal (CSEIJ)*, Vol. 1, No. 5, December 2011, pages 1–10, (2011).
3. S.Abdalla, N. Al-Aama Maryam, A.Al-Ghamdi: Development of MRI Brain Image Segmentation technique with Pixel Connectivity, *International Journal of Scientific and Research Publications*, Volume 6, Issue-6 (2016).
4. Ibtihal D. Mustafa, Mawai A. Hassan: A Comparison between Different Segmentation Techniques used in Medical Imaging, *American Journal of Biomedical Engineering*, 6(2) 59–69 (2016).
5. Ch Murali Krishna, Y. Srinivas: Unsupervised Image Segmentation Using Truncated Log Normal Distribution, *International Journal of Advanced Research in Computer Science and Software Engineering*, Volume 5, Issue 1, (2015).

6. T.V. Madhusudhana Rao, S. Pallam Setty, Y. Srinivas: Content Based Image Retrievals for Brain Related Disease, *International Journal of Computer Applications*, Volume 85, No, 11, pp. 0975–8887 (2014).

7. Anamika Ahirwar: Study of Techniques used for Medical Image Segmentation and Computation of Statistical Test for Region classification of MR Brain Images, I.J. Information Technology and Computer Science, 05, 44–53, MECS (2013).

8. Q. Mahmood, A. Chodorowski, M. Persson: Automated MRI brain tissue segmentation based on mean shift and Fuzzy c-means using a priori tissue probability maps, IRBM 36, 185–196, *Science Direct* (2015).

9. Daniel Biediger, Christophe Collet and Jean-Paul Armspach: Multiple Sclerosis lesion detection with local multimodal Markovian analysis and cellular automata 'growcut", *Journal of Computational Surgery, Springer*, 1:3, pp: 1–15.11 (2014).

10. F Rodrigo, M. Filipuzzi, R. Isoardi, M. Noceti, JP Graffigna: High intensity region segmentation in MR imaging of multiple sclerosis, *Journal of Physics: Conference series* 477, 012024, (2013).

11. B.R. Sajja, S. Datta, R. He, M. Mehta, R.K. Gupta, J.S. Wolinsky, P.A. Narayana: Unified approach for multiple sclerosis lesion segmentation on brain MRI, *Ann. Biomed. Eng.* 34 (1), 142–151, (2006).

12. Saurabh Shah, N.C. Chauhn: Classification of brain MRI Images using Computational Intelligent Techniques, *International Journal of Computer Applications (0975-8887)* Volume-124, No. 14, (2015).

Toward Segmentation of Images Based on Non-Normal Mixture Models Based on Bivariate Skew Distribution

Kakollu Vanitha and P. Chandrasekhar Reddy

Abstract Image analysis mainly focused on identifying the inherent features inside the image for effective understanding of the images. Image segmentation is an integral part of image analysis where, we try to cluster the data and identify meaningful patterns. In this article, we focus upon presenting a model for effective segmentation using non-normal mixture models. The methodology is tested on various image datasets like medical images, natural images, and birds and animals and the result showcases that the model is exhibiting accuracy about 85%, in case of all the images. The performance evaluation carried out using metrics such as image fidelity (IF), peak signal-to-noise ratio (PSNR), and mean squared error (MSE) supports the argument.

Keywords Bivariate skew distribution · Image segmentation · Quality metrics · Classifier accuracy · Performance evaluation

1 Introduction

Image segmentation is considered to be an area in image processing. Segmentation helps to divide the image into constant parts and understand the internal properties within each of these regions [1]. Many models have been developed in the literature for effective segmentation on the images and analysing the features from each of

K. Vanitha (✉)
Department of Computer Science, GITAM University, Visakhapatnam, India
e-mail: vanithagitam@gmail.com

P. Chandrasekhar Reddy
Department of Electronics and Communications Engineering, JNTU, Hyderabad, India

© Springer Nature Singapore Pte Ltd. 2017
S. Patnaik and F. Popentiu-Vladicescu (eds.), *Recent Developments in Intelligent Computing, Communication and Devices*, Advances in Intelligent Systems and Computing 555, DOI 10.1007/978-981-10-3779-5_18

these images. Among the model presented, both statistical mixture-based models and non-mixture models are also highlighted in the literature [2–6].

Among the non-mixture models, most of the models are intended towards the development of procedures based on regions of interest, edge-based, watershed-based, graph cut, saddle point, artificial neural network, etc. [7, 8]; however, it is assumed that these models fail to perform this segmentation accurately when compared to statistical mixture model [9]. This is due to the fact that the statistical mixture models consider the parameters inside the image region, and based on these parameters, the segmentation is carried out [10]. Therefore, it assures higher segmentation accuracy when compared to the other models [11, 12]. With this approximation, many models have been evaluated with the consideration of statistical mixture models. Among these mixture models, some models are based on normality; i.e., Gaussian mixture models are mostly considered in the literature, the basic assumption that every image is considered as normal in shape, and with this assumption, several models have been developed. However, in reality the shapes of the images are mostly non-normal, and hence, it is needed to consider algorithms based on non-normal mixture models for effective segmentation [13–15]. So in this chapter, we therefore have to consider bivariate skew Gaussian distribution for the image segmentation [16]. For accurate result, in this chapter I considered capability metrics such as image fidelity, MSE, and peak signal-to-noise ratio. The outputs derived and compared to those of generalized GMM and the result showcase the performance of the model.

The rest of the chapter is presented as follows: the bivariate skew Gaussian mixture models are explained in Sect. 2. In Sect. 3, the K-means algorithm is presented for effective segmentation, the dataset considered is presented in Sect. 4, and in Sect. 5, the feature extraction is presented. The experimental result is presented in Sect. 6 together with the performance evaluation, and finally, Sect. 7 concludes this chapter.

2 Bivariate Skew Gaussian Mixture Model

This model is used to segment medical images with effective manner. Therefore, this bivariate skew Gaussian mixture model is used for effective segmentation of the medical images. Usually, identities are selected as given input to the model, and the outputs derived will help to cluster the image data into different regions.

In this function, the pixel X follows a multivariate skew normal distribution and the density function of the probability is as follows

$$f(x; \mu, \sigma, \delta) = 2\varphi p(x; \mu, \sigma)\theta 1^3 \delta T \sigma - 1(x - \mu); \quad 0, 1 - \delta \text{ To } \sigma - \delta \quad (1)$$

Three vectors are as follows: one location vector is μ, a scale matrix is σ, and skewness vector is δ.

Let $\varphi p(.; \mu, \sigma)$ be the p-variate density normal distribution with mean vector μ, σ as covariance matrix, and $\theta p(.; \mu, \sigma)$ as the corresponding distribution function.

3 k-Means Algorithm

The k-means algorithm takes the input parameter, k, and partitions a set of n objects into k clusters.

1. Draw a scatter plot and identify the number of clusters k.
2. Randomly choose the cluster mean as one of the data objects.
3. For each of the remaining object, an object is assigned to the cluster to which it is similar.
4. It is based on the distance between the object and the cluster mean.
5. Then, it computes the new mean for each cluster.
6. This iterative process runs until the criterion function converges.

4 Data Set Considered

In order to present the model, we have to considered the images from standard dataset, namely Barkley, UCI, and also some realistic images. The dataset consists of 1000 images taken into account for trainee, and 100 images are tested. Each image is of similar size, and normalization process is applied to normalize the image as illustrated in Table 1.

The outputs derived are represented in the form of the Graphs 1, 2, and 3. From the above graphs, the present developed method performs more effectively, and it gives much more accuracy than the previous GMM.

5 Feature Selection

In order to identify the process of segmentation, the bases of segmentation, i.e. the features, are to be identified approximately. Features play a vital role in the segmentation process. The various features considered are intensity and colour. These features are given as input to the model prescribed in Sect. 2 of this chapter.

Table 1 Segmented images against the performance evaluation

Image	Quality metrics	GMM	BSGMM
	IF	0.321	0.878
	MSE	0.021	0.081
	SNR	12.21	36.15
	IF	0.432	0.898
	MSE	0.231	0.321
	SNR	13.32	36.87
	IF	0.32	0.890
	MSE	0.31	0.321
	SNR	188	37.98
	IF	0.212	0.892
	MSE	0.245	0.1192
	SNR	21.42	37.41
	IF	0.391	0.876
	MSE	0.2514	0.1759
	SNR	3.241	5.68
	IF	0.2134	0.791
	MSE	0.06	0.594
	SNR	13.43	20.39
	IF	0.233	0.923
	MSE	0.01	0.119
	SNR	11.11	29.86
	IF	0.293	0.791
	MSE	0.18	0.213
	SNR	21.214	99

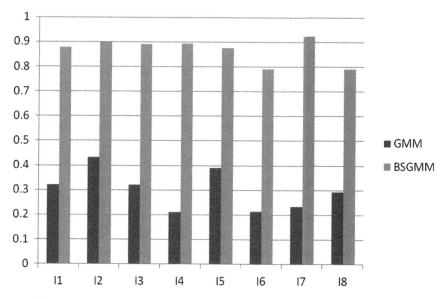

Graph 1 The performance of both the models based on the image fidelity

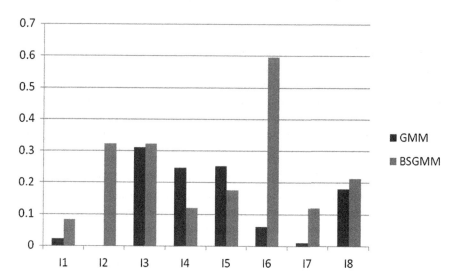

Graph 2 The performance of both the models based on the mean-squared error

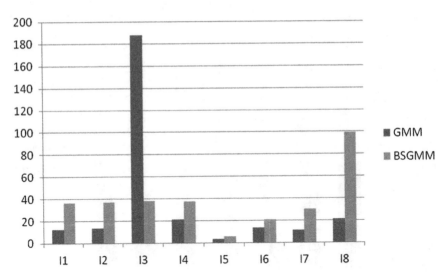

Graph 3 The performance of both the models based on the signal-to-noise ratio

6 Experiment Result and Performance Evolution

In order to present the model, the experiment is carried out in MATLAB environment. Each of the images is considered, preprocessed, and given as input to the *K*-means algorithm for segmentation. From the segmented regions, in order to ascertain the most effected segmentation process, each of the pixels from the image is considered, and the probabilities are estimated using the model presented in Sect. 2. The pixels in the clustered regions are also given to the model, and the PDF is estimated. Based on the log-likelihood estimates, the probability of particular group cab is estimated based on the probability density function. That is, PDF images are considered along with likelihood of the pixels that follow a particular region, comparing the pixels with PDF in segmentation. Once the clustering process is completed, the effectiveness of model is carried out using the performance metrics such as signal-to-noise ratio, image fidelity, and mean-squared error, and the formulas for calculation of each of this above metrics are presented in Table 2.

The evaluation is carried out using the above-mentioned performance metrics, and it is compared with the previous model based on GMM.

Table 2 Formulae for evaluating quality metrics used

Quality metrics	Formula to evaluate
Image fidelity	$1 - \left[\sum\limits_{j=1}^{M} \sum\limits_{k=1}^{N} \left[F(j,k) - \hat{F}(j,k) \right] / \sum\limits_{j=1}^{M} \sum\limits_{k=1}^{N} [F(j,k)]^2 \right]$ where M, N are image matrix rows and columns, respectively
Mean-squared error	$\frac{1}{MN} \left[\sum\limits_{j=1}^{M} \sum\limits_{k=1}^{N} \left[O\{F(j,k)\} - O\{\hat{F}(j,k)\} \right]^2 / \sum\limits_{j=1}^{M} \sum\limits_{k=1}^{N} [O\{F(j,k)\}]^2 \right]$ where M, N are image matrix rows and columns, respectively
Signal-to-noise ratio	$20 \log_{10} \left(\frac{\text{MAX}_I}{\sqrt{\text{MSE}}} \right)$ where MAX_I is maximum possible pixel value of image, and MSE is the mean-squared error

7 Conclusion

In this chapter, a model is developed based on bivariate skew Gaussian distribution for effective image segmentation. In order to display the result in the effectiveness of model, we have considered different images from different benchmark datasets, namely Barkley, UCI, and other real-time images and tested the model. The model has displayed better accuracy during the segmentation process, and the results were distinct with the model based on GMM. The overall accuracy of the model is above 85%.

References

1. Herng-Hua Chang, Daniel J. Valentino, Gary R. Duckwiler and Arthur W. Toga. (2007): *"Segmentation of Brain MR Images Using a Charged Fluid Model"*, IEEE Transactions on Biomedical Engineering, Vol. 54, No. 10, pp. 1798–1813.
2. Adelino R. Ferreira da Silva. (2009): *"Bayesian mixture models of variable dimension for image segmentation"*, Computer methods and programs in biomedicine, 1–14.
3. Ahmet M. Eskicioglu and Paul S. Fisher. (1995): *"Image Quality Measures and Their Performance"*, IEEE Transactions on Communications, Vol. 43, No. 12, pp. 2959–2965.
4. D. L. Pham, C. Y. Xu, and J. L. Prince. (2000): *"A survey of current methodism medical image segmentation,"* Annu. Rev. Biomed. Eng., vol. 2, pp. 315–337.
5. David W. Shattuck, Gautam Prasada, Mubeena Mirzaa, Katherine L. Narra and Arthur W. Togaa. (2009): *"Online resource for validation of brain segmentation methods"*, NeuroImage Volume 45, Issue 2, 1, Pages 431–439.
6. Dr. Samir Kumar Bandhyopadhyay, Tuhin Utsab Paul. (2012): *"Segmentation of Brain MRI Image – A Review"*, International Journal of Advanced Research in Computer Science and Software Engineering, Vol. 2, Issue 3, pp. 409–413.
7. G. Dugas-Phocion, M. Á. González Ballester, G. Malandain, C. Lebrunand N. Ayache. (2004): *"Improved EM-based tissue segmentation and partial volume effect quantification in multi-sequence brain MRI,"* in Int. Conf. Med. Image Comput. Comput. Assist. Int. (MICCAI), pp. 26–33.

8. Guang Jian Tian, Yong Xia, Yanning Zhang, Dagan Feng. (2011): *"Hybrid Genetic and Variational Expectation-Maximization Algorithm for Gaussian-Mixture-Model-Based Brain MR Image Segmentation"*, IEEE Transactions on Information Technology in Biomedicine, VOL. 15, NO. 3.
9. N.R. Pal and S.K. Pal, (1993): *"A review on image segmentation techniques"*, Pattern recognition, vol. 26, no. 9, pp. 1227–1294.
10. J. Ashburner and K.J. Friston. (1997): *"Multimodal image coregistration and partitioning – A unified framework"*, NeuroImage, Vol. 6, pp 209–217.
11. Nagesh Vadaparthi, Srinivas Yerramalle, Suresh Varma Penumatsa, and P.S.R. Murthy, (2011): *"Segmentation of Brain MR Images based on finite skew caussaian mixture models with Fuzzy c-means clustering and EM-algorithm"*, IJCA, vol. 28(10): pp 18–26.
12. Nagesh Vadaparthi, Srinivas Yerramalle, and Suresh Varma. P. (2011): *"Unsupervised Medical Image Segmentation on Brain MRI images using Skew Gaussian Distribution"*, IEEE – ICRTIT+, pp. 1293–1297.
13. Juin-Der Lee. (2009): *"MR Image Segmentation Using a Power Transformation Approach"*, IEEE Transactions on Medical Imaging, Vol. 28, No. 6, pp. 894–905.
14. K. Van Leemput, F. Maes, D. Vandermeulen, and P. Suetens. (2003): *"A unifying framework for partial volume segmentation of brainMRimages"*, IEEE Trans. Med. Imag., vol. 22, no. 1, pp. 105–119.
15. K. Van Leemput, F. Maes, D. Vandeurmeulen, and P. Suetens, (1999): *"Automated model-based tissue classification of MR images of the brain"*, IEEE Trans. Med. Imag., vol. 18, no. 10, pp. 897–908.
16. J.A. Hartigan. (1975): *Clustering Algorithms*, New York: Wiley.

Development of Video Surveillance System in All-Black Environment Based on Infrared Laser Light

Wen-feng Li, Bo Zhang, M.T.E. Kahn, Meng-yuan Su, Xian-yu Qiu and Ya-ge Guo

Abstract During the night or in a dark environment, the definition of current video surveillance systems is very poor, mainly due to the insufficient brightness. Although the infrared LED lights can effectively improve the quality of the picture, the red explosion phenomenon detracts from the stealth of the whole monitoring system and the LED brightness will also decrease after the long-term use. This paper proposes a solution based on an ARM control module platform, a laser light source, and an HD MCCD camera to improve the quality of video surveillance system in the full-dark environment. The new platform occupies a small volume, and the infrared laser light module compared with an LED of the same brightness has low power consumption, no red explosion phenomenon, and reduced heat associated with it. The ARM platform simplifies the video capture and has better image processing capability, which enhances the video quality. This improved the overall versatility of the monitoring system. The video data is stored in a removable SD card format and may be uploaded to the server by means of a wireless network.

Keywords Infrared laser aiding light · Full-dark video surveillance · MCCD · ARM

W. Li (✉) · B. Zhang · M. Su · X. Qiu · Y. Guo
Communication and Information Engineering, Xi'an University of Science
and Technology, Yan ta Road No. 58, Xi'an, Shaanxi, China
e-mail: liwenfeng@xust.edu.cn

M.T.E. Kahn
Cape Peninsula University of Technology, Bellville, South Africa

© Springer Nature Singapore Pte Ltd. 2017
S. Patnaik and F. Popentiu-Vladicescu (eds.), *Recent Developments in Intelligent Computing, Communication and Devices*, Advances in Intelligent Systems and Computing 555, DOI 10.1007/978-981-10-3779-5_19

1   Introduction

Surveillance cameras in the modern society play a more and more important role in enterprises, which can collect high-definition video information of the scene under the low light of the mine or metro video surveillance, video surveillance of the public railway tunnel, and monitoring of the UAV at night. At present, the domestic low-light video surveillance uses LED array infrared using low-light camera. But the power consumption and volume of LED light module are not suitable for portable unit, and IR LED has red burst phenomenon, which affect the concealment of the whole system. Overall, the development trends have a few points. First is the high-definition video, in which the picture quality requirements are getting higher and higher, which require bright IR laser aiding light; second is the intelligence, which is the inevitable trend of the digitalization and network, to achieve early warning and to truly ensure safety; third is the standardization, with the development of monitoring system from analog to digital, which needs to open standards between equipment and system to achieve better connection [1].

Therefore, the new video surveillance system with fully functional, low power consumption, small size, and stable performance can be used in low light and even in the full-dark environment for the rescue communication equipment, which will give the great social benefits. In this paper, the full-dark environment or underground night vision system is proposed based on the ARM embedded system combined with IR laser aiding light module and MCCD camera, which gives more capability and flexibility to process the image.

2   Design of Video Capture System

The night vision video capture system consists of MCCD camera, IR laser light module, ARM embedded control module, power management module, and display module. In the full-dark environment, any low-light camera cannot capture the video or picture without the IR laser aiding light because of no reflection from objects. For improving picture quality, in this system, the MCCD low-light camera is introduced, which has 1/2 supersensitive array, high dynamic range, minimum illumination to 0.0011 lx, and better response than CCD and CMOS in 620–940 nm for hidden shooting. After testing, the higher picture quality would be always presented with the same camera using higher IR auxiliary light power [2]. So the IR laser aiding light power in unit area becomes critical parameter to affect the output signal of the MCCD and picture quality as well.

2.1 IR Laser Aiding Light Module

Currently, the IR LED light array is a most popular light aiding module on market, but the photoelectric conversion efficiency of LED is very poor, so in order to improve the optical power, a large number of infrared LEDs are required, which cause scattering problem and large space for light house. On the other hand, IR LED has brightness attenuation problem, which reduced the system service life and the late picture quality. In this system, the 1-W IR laser aiding light module is proposed with regard to the problems mentioned. Compared with the IR LED source, the IR laser has higher light output with the same power consumption, smaller volume, and longer working life. The constant current source drive circuit of laser is more complex than LED drive circuit, which is shown in Fig. 1. The output current can be adjusted by the program to adapt the different lasers.

The combination of astigmatism lens and convex lens is used to extend the optical path of IR laser light source. The calculation method of illuminated area is shown in Fig. 2. Due to the different lens combinations, the balance between light distance and area could be adjusted. The IR laser aiding light is driven by light detector with IC PAM2803 switched, when the visible light illumination is less than 0.001 lx/F1.2.

2.2 Photoelectric Sensor Module

The MCCD camera is introduced as a photoelectric sensor, which has better performance than CCD or COMS in the low light and wider IR light range. MCCD use

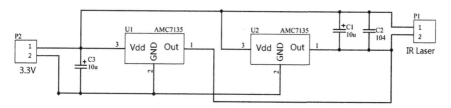

Fig. 1 Drive circuit of IR laser light source

Fig. 2 IR laser light source astigmatism

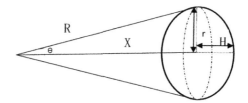

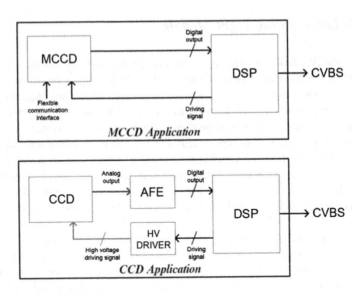

Fig. 3 Comparison of MCCD and CCD circuit

CMYG mutual complementary filter array, which is the same as the CCD passive devices and for greater dynamic range it uses CCD devices, compatible with exposure and readout control. The advanced image signal processor and AFE function are integrated into the MCCD chip like the CMOS high integration characteristics to effectively improve the quality of the image [3]. Because of that, the MCCD could be quantified directly to the output of the digital image signal and without high pressure of CCD driving, so besides saving high-voltage driver IC, power supply is also simplified, as shown in Fig. 3.

2.3 Central Control Module

In this system, the control module uses the ARM platform, which gives many advantages such as a variety of interfaces, humanized operating system interface, and late functional development. The CPU is IC S5PV210, which has clocked up to 1 GHz, with a 32/64 bit of the internal bus structure, 32/32 KB data instruction level 1 cache and 512 KB level 2 cache, and with 2 million instructions per second high-performance computing power [4]. It also has rich peripheral hardware devices, such as camera interface, 24bit TFT real color LCD controller, system power supply controller, serial port, 24-bit channel DMA, three groups of I2C and I2S buses, and USB interface. The whole video capture system is shown in Fig. 4,

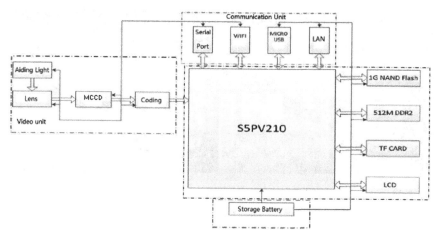

Fig. 4 Video capture system

which includes MCCD camera, IR laser aiding light, TF memory, Wi-fi, and display module [5].

3 System Testing

After the completion of the PCB system board, the printed board would be checked carefully through line by line using the multi-meter, microscopes, and other tools, to prevent the emergence of open-circuit and short-circuit phenomenon. The welding is processed and tested at the same time in order to find and solve the problem as soon as possible.

In order to achieve the system design of the Flash NAND start, the start-up code is written to the Flash NAND storage chip through the Flash NAND boot system. The design uses TF card guide mode, and Flash NAND Program burn and write. Through the Flash NAND boot system, the Android operating system video code software is burn-in. The firmware program provided by the IC company can support the multi-format video code, which is eliminating the complex software programming of the video code algorithm.

As shown in Fig. 5, the picture is displayed in the high resolution of 640 × 480 screen under all-dark work environment around 0.001 lx without IR adding light source and with adding light source, which uses processor s5pv210 H.264 compression encoding and decoding.

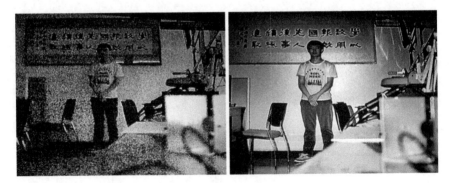

Fig. 5 Picture quality with or without IR aiding light

4   Conclusion

In this paper, the night vision equipment based on IR laser light, MCCD camera, and ARM platform is introduced, which has better performance than LED- and DSP-based night vision systems. This video capture system could be made hand-held information recorder equipment, which can be used in real-time monitoring of underground mine, safety inspection, and emergency communications. Final equipment has a clear image quality, portable, video data to achieve the wireless transmission, and real-time recording, suitable for underground all-dark environment (0 lx). The system is designed using the processor with the H.264 codec processing capacity, which uses 940 nm IR laser light source as an auxiliary lighting and feed through a 640 × 480 resolution TFT-LCD high screen display with low-voltage (less than or equal to 4.2 V), and which ultimately reduces the size of the battery device and equipment. Through the noise reduction circuit and image enhancement technology, the improvement in picture quality would be focused in the next level research.

Acknowledgements This project is funded by Shaanxi Province Science and Technology Innovation Project (2015KTCQ03-10), Xi'An City Research Collaborative Innovation Program (CXY1519(5)), and Xi'An Beilin District Science and Technology Project (Gx1601); scientific research program is funded by Shaanxi Provincial Education Commission (Program NO. 2013JK1079).

References

1. National documentation Standardization Technical Committee.GB3836.4 - 2000 explosion proof electrical equipment for explosive environment. Intrinsically safe electrical equipment "I". Beijing: China Standard Press, 2000.
2. Chu-min Chen, MCCD and CCD, CMOS technology comparison, China's security, Security China & Protection, 2013.04.

3. Ye Junhua. Design of embedded video processing terminal based on ARM11 [D]. Nanjing: Central South University, 2009.
4. Liu Yumin, Li Jianqin, Yao Bin. Development and application of digital video and audio technology. Beijing: National Defense Industry Press, 2003.
5. Li Wenfeng, Gao Jie, Bai Peng. Mine multimedia emergency communication system. 2007 International Conference on Wireless Communications, Networking and Mobile Computing, WiCOM 2007, 2007, p 2865–2868.

Author Biography

Prof. Li was born in November 1969; he is a professor of the School of Communication and Information Engineering, Xi'an University of science and technology; he is a master tutor at the Northwestern Polytechnical College of Communication and System Engineering; he is a professional postdoctoral; and in 2004, he obtained circuits and systems professional doctorate from the Northwestern Polytechnical University. His research directions are the emergency communication technology, the mine rescue communication technology, and the digital signal processing.

Data Preprocessing Techniques for Research Performance Analysis

Fatin Shahirah Zulkepli, Roliana Ibrahim and Faisal Saeed

Abstract Business intelligence (BI) system mixes operational data with the analytical tools to represent descriptive and complicated data to groups of decision makers. BI aims to enhance the features and accuracy of data warehouse to the decision-making process and widely applied in industry. In order to achieve that, BI pulls and gathers information from multiple sources of information systems. Data from multiple sources tend to have flaws such as missing values, inconsistency data, and redundant data. Hence, this paper aims to show data preprocessing techniques used to produce clean and quality data for Universiti Teknologi Malaysia (UTM) research performance analysis. For this research study, required data were provided by UTM management level. In future, this study is expected to compare different data preprocessing techniques and recommend the best one for research performance analysis.

Keywords Business intelligence · Data preprocessing · Research performance

1 Introduction

Business intelligence (BI) offers huge potential for business institution to gain insights into their daily operation, as well as longer threats and term opportunities. BI helps in improving the decision-making process and enhancing the operational efficiency. Becoming a prestige or world-class university is a vital objective that has been set by Institutions of Higher Learning (IHLs) in Malaysia. UTM was

F.S. Zulkepli (✉) · R. Ibrahim · F. Saeed
Information System Department, Faculty of Computing, Universiti Teknologi Malaysia,
Johor Bahru, Malaysia
e-mail: fshahirah78@gmail.com

R. Ibrahim
e-mail: roliana@utm.my

F. Saeed
e-mail: faisalsaeed@utm.my

© Springer Nature Singapore Pte Ltd. 2017
S. Patnaik and F. Popentiu-Vladicescu (eds.), *Recent Developments in Intelligent Computing, Communication and Devices*, Advances in Intelligent Systems and Computing 555, DOI 10.1007/978-981-10-3779-5_20

appointed as one of leading Research University in Malaysia. University rankings reflect reputation or prestige of IHLs, and it plays as a benchmarking tool in monitoring university's performance. This enforce universities to plan strategic course of actions for improvement in quality of higher education environment and to develop capacity of competing in the global tertiary education through acquisition and creation of advance knowledge [1].

Key Performance Indicator (KPI) or known as Key Amal Indicator (KAI) in UTM is created with appropriate metrics or measurable targets to improve the overall university's performance [2]. One of the research strategies suggested is through research performance analysis which has been suggested by [2]. Knowledge Discovery in Databases (KDD) is the process of discovering valuable patterns or knowledge from data sources. Data mining (DM) is part of the whole process of KDD. Data preprocessing is a vital phase in data mining. Data preprocessing is able to solve multiple types of problems in a huge dataset in order to produce a good data [3].

Hence, this paper is structured as follows: The first section is discussed about a brief literature review regarding business intelligence and data preprocessing. Then, the current research approaches in this research study were presented, and the last section will concludes this paper.

2 Literature Review

2.1 Business Intelligence

BI is a system that mixes the operational data with analytical tools to represent complicated and descriptive data to groups of decision [4]. Solomon [3] stated that BI is used to grasp the capabilities available in a business institution: future directions and trends in the markets, state of the business institution, environment which business institutions operating and the technologies. In addition, BI comes with lots of benefits including improving and fastening duration of decision making, increasing operational efficiency, optimizing internal business processes, perfecting flaws, and driving new revenues [5]. In this research study, BI is defined as follows: BI is a method of transforming data into information and from information into knowledge.

The term knowledge in this research study is basically about client's decision-making processes, the competitors, client's needs, and cultural trends; environment of business institutions; and also financial status of a business institution. The idea of this definition is that BI systems ensure desirable message delivered at the right location, at the right time, and in the right form to assist decision makers. As mentioned before, the objective of BI is to enhance the flow and quality of inputs to the decision process, hence facilitating top management business institution.

According to [6], important elements of proactive BI are real-time data warehouse, data mining, exception detection and automated anomaly, automatic learning, and refinement and seamless workflow. Data warehouse (DW) is a data storage which compiled a broad range of data from multiple sources within a company. These data will be used as a guide for management decision. The construction of data warehouse involves data preprocessing techniques. DW is categorized as subject-oriented because DW revolving around big data. DW is considered as not dynamic, time variant, and integrated.

2.2 Data Preprocessing

Data preprocessing is one of the vital steps in data mining. It is used to manage various types of problems in a huge dataset to produce quality data. To proceed with the mining process, the required data should undergo data preprocessing techniques to ensure data are all fit, applicable, and clean. In order to increase the accuracy of the decision making, the quality of the model depends on mining process. Hence, data preprocessing is considered as the key to the problem-solving [7]. Adequate data preprocessing techniques generate a more accurate model and save time. Normally, data preprocessing consists of four steps, namely data cleaning, integration, transformation, and reduction [8]. Data cleaning is used to correct or remove data by filling in missing values, identifying or removing outlier data, smoothing noisy data, and resolving inconsistencies. Data integration aims to combine various sources into one data warehouse. Data transformation is a process that transforms the data into a form that is appropriate for specific data mining algorithms, while data reduction is obtained by reduced representation in volume but produces the same or similar analytical results. All these four steps are important to produce clean and quality data.

3 Research Performance Analysis

UTM is one of the universities in Malaysia that has Research University (RU) status. RU status was obtained by performing great achievements in research and publication and also by achieving KAI (Key Amal Indicators) or generally known as Key Performance Indicators (KPI) set by Ministry of Higher Education (MOHE), Malaysia. To maintain the RU status, the top management level (*decision makers*) request a model and conceptual database which can display and analyze the trend of academic and research publications in UTM (business institution). Each UTM academic staff owns a data dictionary that contains their personal data and job performance. Each of these data records was kept in different data sources; hence, whenever the management level request to look multiple data from various data sources, the process to retrieve the required data is time-consuming. From the perspective of research and performance analysis, a business intelligence dashboard is needed to monitor research accomplishment and performance analysis.

By applying business intelligence, especially in research performance field, the data will be shared effectively and efficiently with top management level.

First and foremost, to create a data warehouse for UTM research performance analysis, the decision makers already listed down their requirements, and from the user requirements, a business insight was created so we can visualize and understand clearly the users' demand. Business insight is a face, a thought, a mixture of data or analysis that includes context and understanding of a situation or issues that show potential for enhancing a business institution. In this phase, we collect data from multiple data sources such as Research and Development Information System (RADIS), Research Management Centre (RMC), Perpustakaan Sultanah Zanariah (PSZ), and other data sources. All the data obtained are considered incomplete as they are lacking attribute values, containing errors and outliers, and inconsistent. We obtained 12 sets of data from the sources, and the data's range is from the year 2010 until the year 2015.

3.1 Data Preprocessing Approach

All 12 sets of raw data obtained from the sources are kept in one data mart in Excel software for ease of the selection process and the inclusion of the attributes. The 12 sets of raw data are being compared with the existing ERD provided by the UTM management level. These are important to identify the required attributes before proceeding to the data integration steps. All these 12 sets of dataset are centralized into one data warehouse and undergo schema integration by entity identification by detecting and resolving data value conflicts [9]. Next, data cleaning step is a step which mainly focused on filling in missing values, converting nominal to numeric, and correcting inconsistent data [10]. Figure 1 shows briefly the data cleaning steps for attributes Gender.

In order to handle missing data, [11] suggests 3 methods on handling this matter: deletion method, single imputation method, and model-based method. Deletion methods have two methods which are listwise deletion and pairwise deletion. These methods are conventional as this is manually handled by the researchers and reduces N [11]. The single imputation method suggests calculating the mean or mode substitution to replace the missing values. Table 1 shows the results of single imputation method.

N representing the total amount of data. As shown in Table 1, missing data occurs on six over eight predictors with grade, faculty name, article in Scopus, Web

Fig. 1 Data cleaning for gender attributes

Gender = M $\rightarrow$ Gender_0_1 = 0

Gender = F $\rightarrow$ Gender_0_1 = 1

Table 1 Variable descriptions

Variable	Definition	Possible values	M	(SD)	N
Age	Age of the researchers	From 27 until 60	4.057	(0.713)	2340
Gender	Gender of the researchers	0 = M 1 = F	0.558	(0.498)	2340
Grade	Wage level	Different according to age and experience	0.235	(0.370)	1983
Faculty name	Researchers' workplace	Different according to work place	3.443	(1.636)	2011
Article in scopus	Amount of articles published	From 0 to 65	10.586	(1.605)	1981
Web of science			0.442	(0.498)	1520
Book chapter			2.783	(1.919)	1197
Journal article non-citation			3.901	(0.398)	2163

of Science, book chapter, and journal article non-citation. By using these statistics may imply that the study have complete data on about half of the dataset. Based on this research study progress, we had achieved halfway of research milestones which are to investigate requirements and criteria of BI framework and to develop a BI framework for UTM re-search performance analysis.

4 Conclusion

In this paper, we have discussed briefly about business intelligence and data preprocessing approach in our research study. However, the study is still ongoing and some of the data are confidential (e.g., UTM ERD). In the future, this study can be applied in other educational fields.

Acknowledgements This work is supported by the Malaysia Ministry of Higher Education (MOHE) and the Research Management Centre of Universiti Teknologi Malaysia under the Fundamental Research Grant Scheme (Vote No. R.J130000.7828.4F741).

References

1. Han, J. & Kamber, M., (2006). Data Mining: Concepts and Techniques Second., San Francisco, CA: Elsevier Inc.
2. Dhillon, S.K. Ibrahim, R. & Selamat, A., (2013). Strategy Identification For Sustainable Key Performance Indicators Delivery Process For Scholarly Publication and Citation. International Journal of Information Technology & Management. 3(3), pp. 103–113.

3. Negash, Solomon. "Business Intelligence." The communications of the Association for Information Systems 13.1 (2004): 54.
4. Agrawal, Akshat, and Sushil Kumar. "Analysis of Multidimensional Modeling Related To Conceptual Level." Analysis (2015): 119–123.
5. Baina, K., Tata, S., and Benali, K. A Model for Process Service Interaction. In Proceedings 1st Conference on Business Process Management (EindHoven, The Netherlands, 2003).
6. Horkoff, Barone, et al. "Strategic business modeling: representation and reasoning." Software & Systems Modeling 13.3 (2014): 1015–1041.
7. Chou, J.-S. et al., (2014). Machine learning in concrete strength simulations: Multi-nation data analytics. Construction and Building Materials. 73, pp. 771–780.
8. Namdev, N. Agrawal, S. & Silkari, S., (2015). Recent Advancement in Machine Learning Based Internet Traffic Classification. Procedia Computer Science. 60, pp. 784–791.
9. Jared, D., (2014). Big Data, Data Mining, and Machine Learning: Value Creation for Business Leaders and Practitioners, Hoboken, New Jersey: John Wiley & Son, Inc.
10. Liu, B. (University of I., (2011). Data-Centric Systems and Applications Second. S. Ceri & M. J. Carey, eds., Heidelberg: Springer.
11. Therese D. Pigott. A Review of Methods for Missing Data (2001). Educational Research and Evaluation. Vol. 7, No. 4, pp. 353–383.
12. Chong, M., (2005). Traffic accident analysis using machine learning paradigms. Informatica. 29, pp. 89–98.

Author Index

A
Abdul-Majjed, Ielaf Osaamah, 31
Abhijyoti, Ghosh, 113
Ahmad, Asraful Syifaa', 9
Ahmad, Mohamad Nazir, 9
Alzahrani, Salha, 17
Anuradha, S., 133

B
Bai, Yunyue, 1
Banani, Basu, 113

C
Cao, Jianzhong, 129
Chandana, P., 21
Chandrasekhar Reddy, P., 141
Chaudhary, Jyoti, 61

D
Dai, Xinguan, 1
Dakhole, Pravin, 89
Dua, Mohit, 105

E
Eisa, Taiseer, 17

G
Garg, Shruti, 45
Gauthami Latha, A., 21
Guo, Ya-ge, 149

H
Hassan, Rohayanti, 9
Hussain, Jamal, 73

I
Ibrahim, Noraini, 9
Ibrahim, Roliana, 157

J
Jain, S.C., 61
Johari, Punit Kumar, 97

K
Kahn, M.T.E., 149
Khorgade, Manisha, 89

L
Lalmuanawma, Samuel, 73
Li, Wen-feng, 149
Lolit Kumar, Singh L., 113

M
Mahapatra, Sakuntala, 53
Mohanta, Debasis, 53
Mohanty, Prasant Kumar, 53

N
Nayak, Santanu Kumar, 53

P
Patil, Kiran Kumari, 121
Patil, Shantala Devi, 121

Q
Qiu, Xian-yu, 149

R
Ramlan, Rohaizan, 9

S
Saeed, Faisal, 157
Sahu, Priyanka, 105
Salim, Naomie, 17
Sanjay, Ghosh Kumar, 113
Satyanarayana, C.H., 21, 133
Sharma, Arvind K., 61

S. Patnaik and F. Popentiu-Vladicescu (eds.), *Recent Developments in Intelligent Computing, Communication and Devices*, Advances in Intelligent Systems and Computing 555, DOI 10.1007/978-981-10-3779-5

Shrivastava, Sunidhi, 97
Srinivas, Y., 21
Srinivas Rao, P., 21
Su, Meng-yuan, 149
Subhradeep, Chakraborty, 113
Sudipta, Chattopadhyay, 113
Suri, Shelza, 37

T
Tang, Shancheng, 1

V
Vanitha, Kakollu, 141
Vijay, Ritu, 37

Vijayakumar, B.P., 121

W
Wang, Bin, 1
Wu, Di, 129

Y
Yang, Hongtao, 129

Z
Zhang, Bo, 149
Zhou, Zuofeng, 129
Zulkepli, Fatin Shahirah, 157

Printed in the United States
By Bookmasters